Elementary Lessons in Magnetism

by W. Jerome Harrison

with an introduction by Roger Chambers

Self Reliance Books

Get more historic titles on animal and stock breeding, gardening and old fashioned skills by visiting us at:

http://selfreliancebooks.blogspot.com/

Introduction

I am pleased to present yet another title in our "How To ..." series.

The work is in the Public Domain and is re-printed here in accordance with Federal Laws.

As with all reprinted books of this age that are intended to perfectly reproduce the original edition, considerable pains and effort had to be undertaken to correct fading and sometimes outright damage to existing proofs of this title. At times, this task is quite monumental, requiring an almost total "rebuilding" of some pages from digital proofs of multiple copies. Despite this, imperfections still sometimes exist in the final proof and may detract from the visual appearance of the text.

I hope you enjoy reading this book as much as I enjoyed making it available to readers again.

Roger Chambers

PREFACE.

THIS little book is intended strictly as an *introduction* to the science of which it treats. The Education Department requires that the teaching of Elementary Science in the schools under its direction should be " purely descriptive and experimental," and I have here attempted to indicate how such teaching should be carried on. Our *theories* may and do frequently change, but the *facts* of Nature upon which they are founded are immutable. Let us study the facts, and the theories will adjust themselves.

Each and every chapter of this book has been given as an object-lesson many times to classes of children averaging sixty in number, and of ages from ten to sixteen. Every encouragement should be given to young students to experiment on their own account; and it will be found that they are wonderfully eager to do so, and also that many of them are very apt in the construction of simple apparatus. Care has been taken to describe mostly such experiments as any student may repeat at home, with but little expenditure of money.

W. J. H.

BIRMINGHAM, *January 1895.*

CONTENTS.

———◆◆———

APPENDIX.

MAGNETISM.

I.—NATURAL MAGNETS.

1. Discovery of Magnets—2. Derivation of the name "Magnet"—3. Nature of Natural Magnets—4. Properties of Natural Magnets—5. Discovery of the "Leading" Power of Natural Magnets—6. Armatures—7. What is Magnetism?

1. Discovery of Magnets.—Some of the properties of magnets have been known for so long a time that we do not know who was the first to notice them; indeed, it is probable that they were discovered, independently, by different people in different places.

Ancient Chinese writings, going back to 1000 B.C., contain the earliest positive mention of magnets; but the Greek and Latin nations of south-eastern Europe were also familiar with the fact that magnets could attract iron, long before the Christian era. The Latin poet Lucretius, who died 55 B.C., writes :—

> " Now, chief of all, the magnet's power I sing,
> And from what laws the attractive functions spring.
> The magnet's name the observing Grecians drew
> From the magnetic region where it grew ;
> Its viewless, potent virtues men surprise,
> Its strange effects they view with wondering eyes,

When, without aid of hinges, links, or springs,
A pendent chain we hold of iron rings
Dropt from the stone—the stone the binding source—
Ring cleaves to ring, and owns magnetic force!"

2. Derivation of the name "Magnet."—Most of the natural magnets which were in the possession of the Greeks came from the neighbourhood of *Magnesia*, a city of Lydia, which is a province in Asia Minor. They were therefore called "Magnes-stones," which ultimately became shortened into "magnets."

An old story attributes the discovery of the magnet to a shepherd of Magnesia, who chanced to leave his iron crook against a rock (which was a natural magnet), and who was surprised on his return to find the crook not fallen to the ground, but suspended from a point of the rock.

3. Nature of Natural Magnets.—A natural magnet is a heavy stone, varying in colour from gray to black, with more or less metallic lustre. Chemical analysis proves that all natural magnets are composed of an oxide of iron, containing three parts of iron to four parts of oxygen. As in chemistry iron is represented by the letters Fe (from *ferrum*, the Latin word for iron), and oxygen by the letter O, we may briefly state the composition of natural magnets by writing it as Fe_3O_4. This substance is known as the "magnetic oxide of iron," and as the "black oxide of iron." There are large deposits of iron ore of this composition in Norway and Sweden, in the island of Elba, in Siberia, and in Arkansas in the United States. Although the chemical composition of the ore is alike, or nearly so, in all these localities, yet pieces having the magnetic properties are rare, so that a small natural magnet costs from two to five shillings. Excellent iron is made from the magnetic iron ore.

4. Properties of Natural Magnets.—How is a natural magnet to be easily distinguished from any other hard black stone?

(1.) In the first place, a stone having magnetic properties can *attract and support small pieces of iron.* If the natural magnet is rolled among a quantity of iron filings, many of the little particles of iron will cling to the stone (Fig. 1), forming tufts at two points on its surface; and if the magnet is lifted up, the filings will still adhere to it, and can only be rubbed off with some difficulty. Small iron nails, too, or iron

FIG. 1.—Natural Magnet surrounded with clusters of iron filings.

rings, will be supported by some power or force possessed by the natural magnet, when they are placed in contact with the points of its surface to which the tufts of filings adhered. If a nail be hung up by a thread, the natural magnet will even draw it towards itself from a distance of an inch or more. Natural magnets, however, have not usually much power of attraction. Sir Isaac Newton possessed a small natural magnet weighing but three grains, which he had set in a finger-ring, and which is said to have been able to lift a weight of iron of 700 grains, or more than two hundred times its own weight. A native magnet sent by the Emperor of China as a present to the King of

Portugal weighed 38 lbs., and was able to sustain 200 lbs. of iron, or about five times its own weight. These two examples will serve to illustrate the well-known fact that the power of magnets does *not* increase in proportion with their weight.

(2.) Now tie a thread round the stone which is believed to be a magnet, and suspend it by the thread so that it can turn freely in any direction. To allow this, the thread should itself be without any tendency to twist or untwist. When so suspended, the stone (if it is indeed a natural magnet) will only remain at rest when the ends to which the filings adhered are *pointing north and south*. If the stone is moved round until it points east and west, and is then released, it will immediately swing round, and will again settle in its north and south position, in which it will remain as long as it is not disturbed. It was this property which obtained for the magnet the name of *lodestone*, or *leadstone*, by which it was, and indeed is still, often called. It is plain that a stone possessing such a property as that of always pointing in a certain direction could lead or guide a person across desert plains or over the sea, when without such an indicator the proper route could not be followed.

5. Discovery of the "Leading" Power of Natural Magnets.—It was not till long after the discovery of the power of a magnet to attract iron that men found out its still more remarkable property of pointing north and south. Again the Chinese seem to have been before the Western nations; for Chinese records show that magnets were used on board their junks to aid in the navigation of the Chinese seas in the year 300 A.D., while even before that time they had employed the lodestone as a guide over the vast trackless plains of Central Asia. In Europe, it was not till the year 1300 that the discovery of this, the most useful property of the magnet, was made by an Italian named Gioja.

6. **Armatures.**—Lodestones are usually rough and irregular in shape. It is advantageous to work them into an oblong form, cutting the stone so that the points where the magnetic power is greatest lie one at each end of the block (as at N and S, Fig. 2).

Each end of the lodestone may then be protected or "armed" by an L shaped piece of soft iron, called an *armature* (as at *a b* and *c d*, Fig. 2). These armatures not only protect the natural magnet from injury, but also allow the two ends of the stone

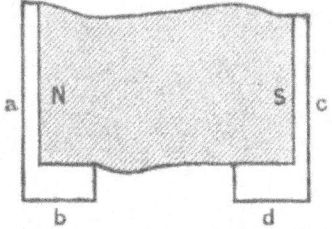

FIG. 2.— Natural Magnet, with its poles marked N and S, and provided with L shaped armatures *a b* and *c d.*

to *act together* when required. For it is found that the magnetic force is transmitted along the soft iron from *a* to *b* and from *c* to *d*; thus the two points *b* and *d* can act together upon any piece of iron which is placed across *b d,* and in this way a much greater weight can be sustained than if either end of the lodestone were acting *by itself.*

This is the original and proper use of the word "armature," but it is now very frequently also used as another name for the "keeper" of a magnet.

7. **What is Magnetism?**—Every lodestone is the seat of a power or force to which the name of *mag-*

netism is applied. This magnetism cannot be any kind of *matter*, for its presence makes no difference in the *weight* of a magnet, and we know that all matter has weight. That magnetism is a *force* is proved by the manner in which it is able to cause small pieces of iron to *move* towards the centres of attraction—the ends of the magnet. The force itself is called magnetism, and to any piece of matter which is able to exert this force we give the name of a *magnet.*

II.—ARTIFICIAL MAGNETS.

8. Difference between Iron and Steel.—Iron is a metallic *element;* therefore *pure* iron contains nothing but iron. Perfectly pure iron is difficult and very expensive to obtain. Ordinary cast or pig iron contains several impurities, from the greater part of which it is freed by the processes of refining and puddling, after which it is known as wrought-iron.

Steel is made by heating wrought-iron in contact with charcoal in a furnace. A small portion of the iron combines with a little of the carbon to form a chemical compound called carbide of iron, and this remains mixed up with the remainder of the iron, to which it imparts certain peculiar properties. The following table shows the difference in the composition of cast-iron, wrought-iron, and steel :—

	Cast-Iron.	Wrought-Iron.	Steel.
Iron..........	93	$99\frac{1}{2}$	$98\frac{1}{4}$
Carbon........	3	$\frac{1}{4}$	$1\frac{1}{2}$
Impurities....	4	$\frac{1}{4}$	$\frac{1}{4}$
	100	100	100

It will be noticed that wrought-iron contains very little carbon—only ¼ lb. in 100 lbs.—while cast-iron has much more—3 lbs. in 100 lbs. Steel takes an intermediate position, containing about 1½ lbs. of carbon in 100 lbs. But it must be remembered that in the case of steel the carbon is *chemically combined* with a portion of the iron, while in wrought and in cast-iron the carbon and the iron are simply *mixed* together.

Pure iron is a comparatively soft, tenacious metal. The Swedish charcoal iron (so called because it is made by the use of wood-charcoal instead of coal in the furnaces) is sufficiently pure for use in experiments in magnetism, while ordinary wrought-iron also answers fairly well.

The most remarkable property of steel is the great hardness which it attains on being plunged into cold water while in a red-hot state. It then becomes so hard as to scratch glass, and is also extremely *brittle*. From this state it can be *tempered* by exposing it to a right amount of heat, according to the purpose for which it is required, and again cooling it quickly.

9. Artificial Magnets.—An artificial magnet is one to which magnetic power has been imparted by the art of man, as distinguished from the natural magnets which obtained their magnetism while lying in the earth.

The production of artificial magnets was a great step in advance, but it was made so long ago that the history of the discovery is unknown. As the method is a very simple one—just rubbing a piece of steel with one end of a lodestone—it is probable that the discovery was accidental. Artificial magnets can be made of any shape we please; and as they are also more powerful than lodestones, they are now invariably used for practical purposes instead of the latter.

There are several varieties of artificial magnets, which may be classified as follows:—

10. Classification of Artificial Magnets.—

A. PERMANENT MAGNETS.
- (1.) Bar Magnets.
- (2.) Horse-shoe Magnets.
- (3.) Magnetic Needles.
- (4.) Compound Magnets, or Magnetic Batteries.

B. TEMPORARY MAGNETS.
- (5.) "Induced" Magnets.
- (6.) Electro-magnets.

11. Permanent Magnets.—

All permanent artificial magnets of good quality consist of pieces of hard-tempered steel to which magnetic power has been imparted by one or other of several methods, which we shall presently describe. Cast-iron, and even common wrought-iron, will retain a certain amount of magnetic force, but far inferior in strength to that possessed by suitably-prepared steel. The best steel for permanent magnets is made in Sheffield, and contains, in addition to its carbon, a small quantity of the metal named tungsten, whence it is known as *tungsten steel*. When a permanent magnet is kept in a proper manner, its magnetic force remains the same, year after year.

12. Temporary Magnets.—

Soft iron is the most suitable substance of which to make a temporary magnet. By "soft" iron is here meant iron which itself becomes endowed with magnetic force when brought near a permanent magnet, but which immediately *loses its magnetism* when it is removed from such a position. Such iron would, however, be also "soft" in the ordinary sense of the term; for it is found that to act in this manner the iron must be pure, or very nearly so, easily worked, and carefully *annealed.* The annealing process consists in making the iron red-hot, and then causing it to cool very slowly and very regularly. Thus, after the soft iron bar has been raised to a red heat, it may be embedded in a heap of hot ashes, or the fire itself may be allowed gradually to die out, and the iron to cool with it, so that the entire cooling process may take twenty-four hours or longer.

13. **Forms of Permanent Magnets.**—The ordinary *Bar Magnets* consist of straight strips of steel, whose length should be about fifteen times and thickness about one-fourth of their breadth. Thus a bar magnet 15 inches long should be about one inch broad and one-fourth inch thick. Such bar magnets are best kept in grooves in a wooden box (see Fig. 3), the ends being connected by short pieces of soft iron called "keepers."

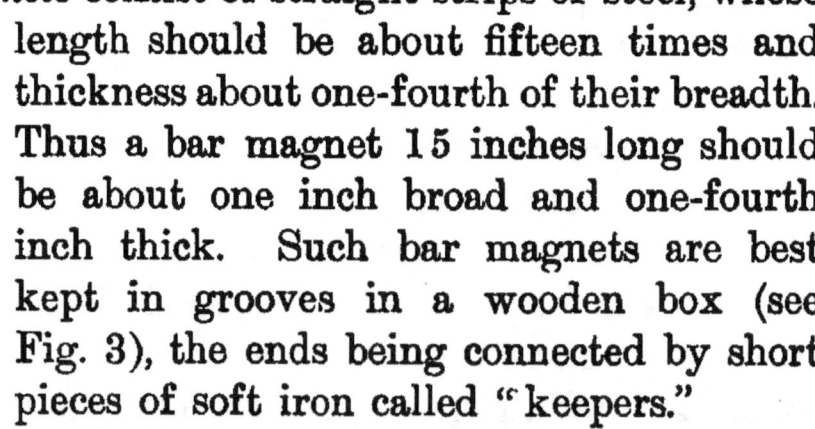

Fig. 3.—Pair of Bar Magnets (with keepers).

Horse-shoe Magnets consist of steel bars (which should have the same proportions as those for bar magnets) first bent into the form of a horse-shoe, and then hardened and magnetized. The advantage of this form of magnet is that the two poles can *act together* in sustaining a weight of iron. In this way a good horse-shoe magnet weighing one pound ought to bear a weight of 20 lbs. hanging from the keeper. When a horse-shoe magnet is laid aside, care should be taken to place a soft iron keeper across the ends, or the magnetic power will slowly diminish (Fig. 4). It is a useful practice also to cover with red varnish (made of sealing-wax dissolved in spirits of wine) all the middle portions of both bar and horse-shoe magnets; this prevents them from rusting. A little oil or grease should be smeared over the bright ends when the magnets are put away.

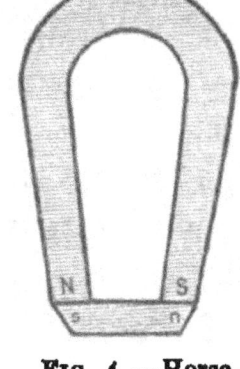

Fig. 4. — Horse-shoe Magnet (with keeper).

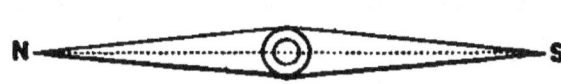

Fig. 5.—Magnetic Needle with Agate Cup (seen from above).

Magnetic Needles are formed of very thin light strips of steel, cut to the shape of a lozenge. A hole is punched in the centre, and into this is fitted a little cup of brass, glass, or agate, by which the needle can be poised on a sharp point (see Figs. 5 and 6). The advantage of this plan is, that a very small force is sufficient to cause such a needle to move, so that it can be employed for delicate experiments.

Fig. 6.—Magnetic Needle on pivot.

Compound Magnets.—The most powerful per-

manent magnets are made by fastening together
a number of magnets made of thin sheet steel, such

as that called steel-
busk, which is used
in staymaking (Fig.
7).   Such a com-

FIG. 7.—Compound Magnet (middle
strip longest).

pound magnet, magnetic battery, or laminated magnet (for it is known by all these names), is several
times stronger than a magnet made of a single bar
of steel of the same weight.   In these compound
magnets the central bar or strip should be the longest,
and those placed on each side of it should each be a
little shorter as we proceed towards the outside of
the row.   It is also advisable to place strips of
card-board between the magnets, so as to separate
the ends, at all events, from one another.

14. Kinds of Temporary Magnets.—To the first
class of temporary magnets we have given the name
of *Induced Magnets*.   These consist of
pieces of soft iron.   When one end of
a piece of soft iron is in contact with a
magnet, the iron becomes itself endowed
with magnetic power, and its other end
is able to attract another piece of soft
iron ; this in its turn is magnetized, and
so on (Fig. 8).   But if the permanent
magnet is taken away, all the pieces of
soft iron lose their magnetism, and the
chain falls to pieces.   It is even sufficient for a piece of soft iron to be *near*

FIG. 8.—Series
of short bars
of soft iron sus-
pended from
Permanent
Magnet.

a powerful magnet, without actually touching it,
for the same effects to be produced.

15. Electro-magnets.—To the second class of temporary magnets the name of *Electro-magnets* is applied, since their magnetism is due to the force of electricity.   If a copper wire, through which a current of electricity is flowing, is coiled round and round a soft iron bar, the iron will become strongly magnetic.   When the copper wire is carefully covered with silk, and wrapped round the soft iron many times, a magnet of great power will be

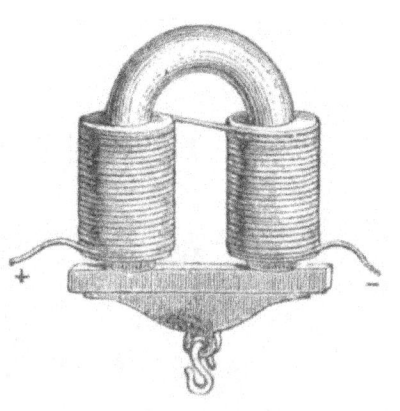

FIG. 9.—Electro-magnet (with keeper).

produced (Fig. 9).   These electro-magnets can be made to far surpass all others in their sustaining power, and have, indeed, been so constructed as to hold up a weight of several tons attached to the keeper.   But directly the current of electricity ceases to flow in the encircling wire, the iron core loses its magnetism, and the keeper falls off.

III.—MAGNETIC INTENSITY.

16. Apparent Distribution of the Magnetic Force in a Magnet.—In the various experiments which we are about to describe, we shall in all cases consider artificial magnets to be used, unless the contrary is stated, since they act in precisely the same manner as natural magnets, being at the same time cheaper, of more convenient shape, and possessing more magnetic force in proportion to their size and weight.

Fig. 10.—Bar Magnet which has been rolled in iron filings.

The simplest experiments will teach us that all parts of a magnet cannot exert the same force upon external substances. When a bar magnet is rolled in a heap of iron filings, the filings adhere in great tufts or clusters to the ends of the magnet, while the middle is left bare (Fig. 10). It is the same with a horse-shoe magnet: a great mass of filings can be raised by the two ends, while to the middle part, or bend, not a single filing will adhere. This clearly proves that, as far as the power to attract external objects is concerned, the magnetic force is greatest near the ends of the magnet, and decreases as we approach a point half-way between the two ends.

To test the matter more exactly, we may procure a number of little cylinders of soft iron—pieces of iron wire half-an-inch long will do—or little iron bullets, or iron rings, and see what number of them can be suspended from the different parts of the magnet. At the extreme end of the magnet we perhaps find that six of our rings can be sustained, one below the other. Half-an-inch from the end the magnetic force is only able to sustain five; going towards the centre of the bar the magnetic force decreases rapidly, and four, three, two, and at last one ring only can be held up; while in the centre, and for some space on each side of it, no ring at all will remain suspended (Fig. 11). When the other end of the magnet is examined, exactly the same effects are

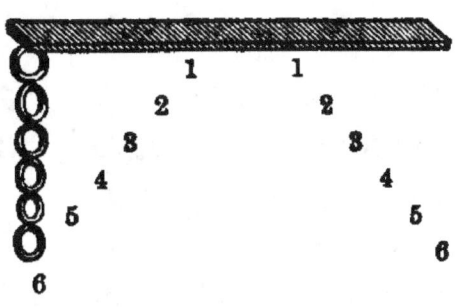

Fig. 11.—Series of iron rings suspended from Magnet.

observed, the magnetism decreasing rapidly as we advance from the end of the bar towards the middle.

The experiments with the filings, and with the bits of iron wire or iron rings, teach us that the magnetic force is greatest at or near the ends of a magnet, and that it decreases rapidly as we approach the middle, where there is no (free) magnetic force at all.

17. **Curve of Magnetic Intensity.**——Having performed the experiment described in the preceding paragraph, we can proceed to draw a figure which will enable us to ascertain the exact amount of magnetic force existing at any point of the length of the

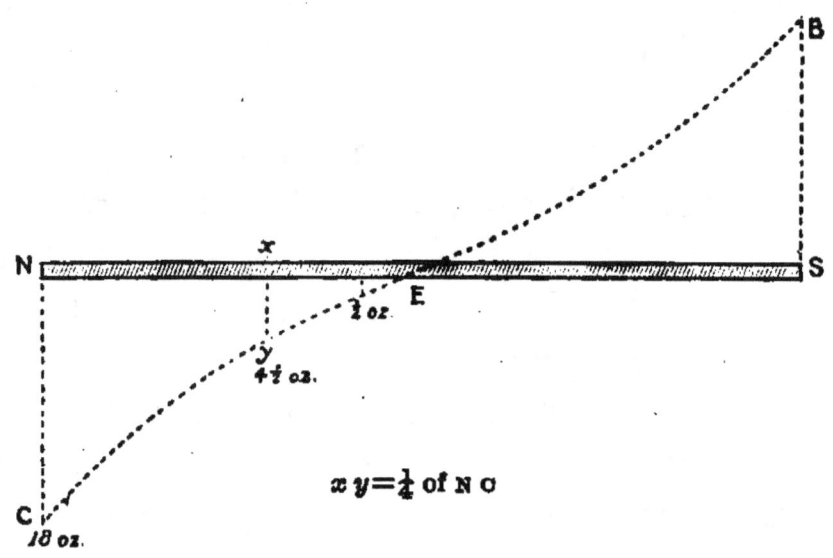

FIG. 12.—Curve of Magnetic Intensity.

magnet. Let N S (Fig. 12) represent a long and narrow bar magnet, and let the vertical straight lines each represent the weight of iron, say the number of ounces, which could be sustained by the magnet at that point, varying from 18 oz. at each extremity to half-an-ounce at a point not far from the centre.

Connect the ends of these vertical lines by a curved line, C E B, which shall touch the magnet at only one point—in the centre. This curved line may be called the curve of magnetic intensity, and the force existing at any point along the magnet may be ascertained by drawing a straight line from that point, at right angles to the magnet, until it meets the curve. Thus, suppose it is required to ascertain, without actual experiment, the magnetic force existing at the point x; draw $x\,y$ at right angles to N S. The length of $x\,y$ is one-fourth of the line N C, which represents a weight of 18 oz.; the magnetic force existing at the point x would therefore be one-fourth of the force existing at N or S, and would consequently be able to sustain $4\frac{1}{2}$ oz. of iron.

18. **The Magnetic Field.**—The influence of a magnet extends over a considerable space all round it, and the stronger the magnet the greater is the distance to which it can make its power felt. To the region over which the influence of a magnet extends the name of the *magnetic field* is given. This region stretches in every direction from the magnet—above and below, as well as on either side. The strength of the magnetic field is greatest in the spaces immediately around each extremity of the magnet, and diminishes steadily as we recede from the magnet in any direction.

19. **Faraday's Curves, or Lines of Force.**—The existence of the magnetic field, and the direction of the magnetic forces in it, may be beautifully shown in the following way. Lay a short powerful bar magnet upon a level surface, cover it with a sheet of cardboard or of glass, and shake upon the card some iron filings, either by means of a pepper-caster, or by sifting them through muslin. Tap the edges

of the card, and the filings will arrange themselves, end to end, in beautiful curves (see Fig. 13), which spread out in a fan-like manner from each end of the magnet, those from the one end curving round on either side to meet those from the other.   The

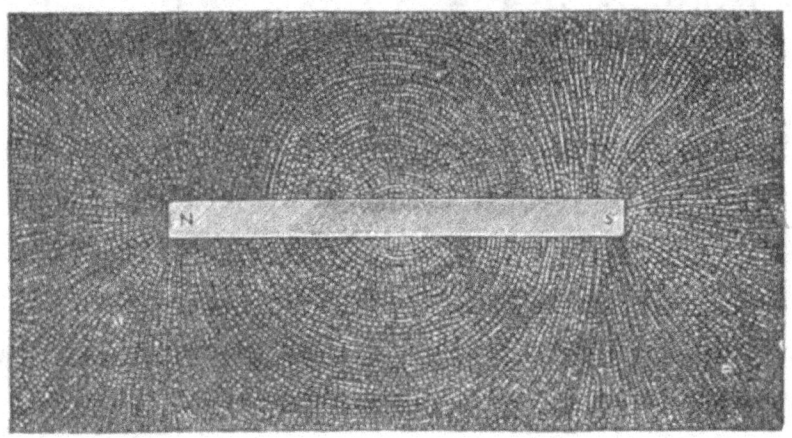

FIG. 13.—Curves of Force round bar magnet.

great philosopher, Michael Faraday, who studied these lines deeply, called them "lines of force," because they indicate, at every point of the magnetic field, the precise direction of the magnetic forces. Every particle of iron present in each curve is itself, for the time being, a perfect little mag-

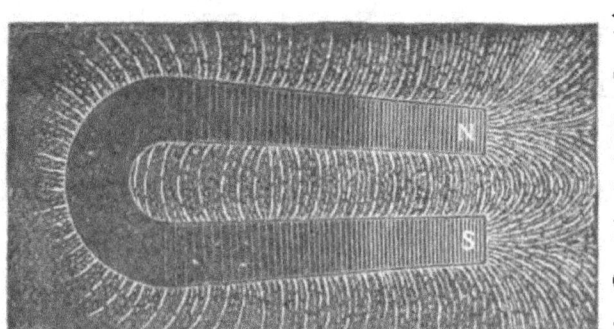

FIG. 14.—Curves of Force, horse-shoe magnet.

net.   The curves produced around the poles of a horse-shoe magnet are shown in Fig. 14.   It will be noticed that the curves are close together in those parts of the magnetic field (near the ends of the magnet) which have the greatest intensity, and that they separate from one another, or widen out, the greater their distance from those points.

These "magnetic figures," as they may be called, may be preserved for future use by first gumming the card and allowing it to dry, then forming the figures, after which a jet of steam is allowed to play on the card. The steam softens the gum, in which the filings embed themselves, and the gum again drying keeps all the filings affixed to the card and in their proper curves. Another method is to brush the card over with melted paraffin wax, which soon solidifies, and on which the curves must be obtained. A hot flat-iron afterwards passed over the card, at a distance of half-an-inch from the surface, will again melt the paraffin, and so fix the filings in their places.

IV.—POLARITY OF MAGNETS.

20. Names of the Parts of a Magnet.—To the ends of a magnet, where the magnetic force appears to be concentrated, the name of *poles* is applied. In ordinary magnets the poles are not quite at the extremities of the magnet, but very near them: in a bar magnet ten inches in length, there would be a pole at a distance of about one inch from each end. The use of the term "poles" for the ends of a magnet arises from the fact that when a magnet is so suspended that it can freely turn in any direction, one *end* of it always points towards the Earth's North *Pole*, and the other *end* towards the Earth's South *Pole*.

If we call the ends of a magnet its *poles*, it is natural to call the middle part, or half-way line, of the magnet its *equator*, since, like the Earth's Equator, it is midway between the two poles. The equator of a magnet is also called the *neutral line*, to express the fact that the magnet is there neutral, or indifferent, to substances (such as iron) which are influenced by its poles.

21. Distinction between the Poles of a Magnet.—Let a number of magnetic needles be balanced on pivots, placed at a distance of several feet apart, so that the magnets cannot affect one another. All these magnetic needles will point towards the Earth's North Pole; and if one end of any magnetic needle be *marked*, as by making a nick with a file upon it, it will be found that it is the *same end* of a magnet which always points towards the Earth's North Pole. If this *north-seeking* end, or pole, of the magnet be turned round by the hand so that it points south-wards, it will not remain in that position, but will swing round into its old place. The other end, or pole, of the same magnet acts, of course, in exactly the opposite manner, so that we term it the *south-seeking* end, or pole. Sometimes the N.-seeking end is called the *red pole*, and the S.-seeking the *blue pole* of the magnet, because it is the custom to paint them of these colours. The N.-seeking end is also called the *marked* pole, because it is usually dis-tinguished either by a nick cut across that end of the magnet, or by the letter N being stamped in the steel.

22. Soft Iron is attracted by either Pole of a Magnet.—*Either* pole of a magnet can attract the same piece of soft iron. Suspend a piece of soft iron by a thread (the French "screw-nails" are usually made of excellent soft iron, and answer well for these experiments), and hold the N.-seeking pole of a magnet within an inch or so of the suspended iron; the iron will quickly move towards the N.-seeking pole. Repeat the experiment with the S.-

seeking pole of the same magnet, and the iron will be attracted with precisely the same force. To vary this experiment, we may place the piece of iron upon a large cork, floating in the centre of a basin of water, and draw it to the side of the basin first with the N.-seeking and afterwards with the S.-seeking pole of the same magnet.

23. Iron also attracts a Magnet.—In all the experiments hitherto described, the iron only has been free to move, while the magnet has been held in the hand. If we now reverse this, and suspend the magnet by a thread, we shall find that a piece of iron will draw, or attract, a magnet in exactly the same manner as the magnet attracts the iron. And when the magnet is floated upon a cork, it will be attracted to the side of a basin when a piece of iron is held there. The fact is, that *each* of the two bodies *attracts the other:* a magnet attracts iron, and the iron also attracts the magnet. If the magnet and the iron be each balanced upon corks placed on water, they *will move towards each other*, and will meet (supposing them to be of the same weight, etc.) exactly half-way.

24. Decrease of Attraction with Increase of Distance. —It is very plain that the attractive power which a magnet exerts on a piece of soft iron decreases very rapidly as the distance of the magnet from the iron is made greater. If a bit of iron is hung up by a thread, and a magnet is gradually brought nearer and nearer to the iron, the latter will move more and more towards the magnet, until, when the distance of the magnet is very small, its attractive power is so great that the bit of iron makes a sudden rush towards it.

25. Law of Inverse Squares.—In what way, and to what extent, does the influence of the magnet decrease as it is removed farther and farther from a piece of iron ? At first sight it seems probable that when the distance between the two bodies is *doubled*, the attraction will be *halved ;* but experiments show that the attractive force falls off much more rapidly than this. The final result of a great number of experiments has been to establish the fact that a law which is known to be true for heat, for light, for sound, and for gravitation, also holds good with regard to magnetism. This law states that "the force of attraction of a magnet acting on a piece of iron is *inversely proportional* to the *square* of the distance between them." This famous expression is often spoken of as the "Law of Inverse Squares." As an example, let us inquire how much less strongly a magnet will attract a piece of iron at a distance of two inches than at a distance of one inch.

In each case square the number representing the distance :—

$$1 \times 1 = 1$$
$$2 \times 2 = 4$$

Now invert the numbers so obtained : since $1 = \frac{1}{1}$, it is clear that in the first case no change is made by this process of inversion ; but $4 = \frac{4}{1}$, and this inverted becomes $\frac{1}{4}$. These two numbers—1 and $\frac{1}{4}$—represent the comparative forces with which the magnet attracts the piece of iron at the distance of one inch and of two inches respectively; and we see that when we *double* the distance the magnetic force is only *one-quarter* as great as before.   So, also, at a distance of three inches the force is only *one-ninth* as great as at a distance of one inch ; at a distance of four inches, only *one-sixteenth* as great, and so on.

If, on the contrary, the magnet be brought *nearer* to the iron—say to the distance of half-an-inch— the attraction will be *increased fourfold ;* at one-third of an inch it becomes nine times as great as at one inch, and so on.

V.—MAGNETIC REPULSION.

26. How a Magnet acts upon another Magnet.—We have learned that *either* pole of a magnet will attract a piece of soft iron. The two poles of the same magnet will even act together in attracting the same piece of soft iron, as we see in the manner in which the keeper of a horse-shoe magnet is retained by the joint action of the two poles.

But how will one magnet act upon another magnet? This question can only be answered by trying the experiment. Bring, therefore, the N.-seeking pole of a small bar magnet near to the S.-seeking pole of a balanced magnetic needle. The two poles attract each other strongly, and the blue end of the needle rushes towards the red end of the bar magnet. So far, the case is exactly like what happened when the magnet was brought near a piece of soft iron.

But now remove the bar magnet, turn it round, and gradually bring its S.-seeking pole near to the S.-seeking pole of the needle. Now we see something quite new—the S.-seeking pole of the needle *moves away* from the S.-seeking pole of the bar magnet, and may be driven round and round as often as we please.   In the case of the action of the magnetic force upon soft iron we got nothing but *attraction;* but we see that when one magnet acts upon another magnet, we can obtain both attraction and *repulsion.*   Repeat the last experiment, now

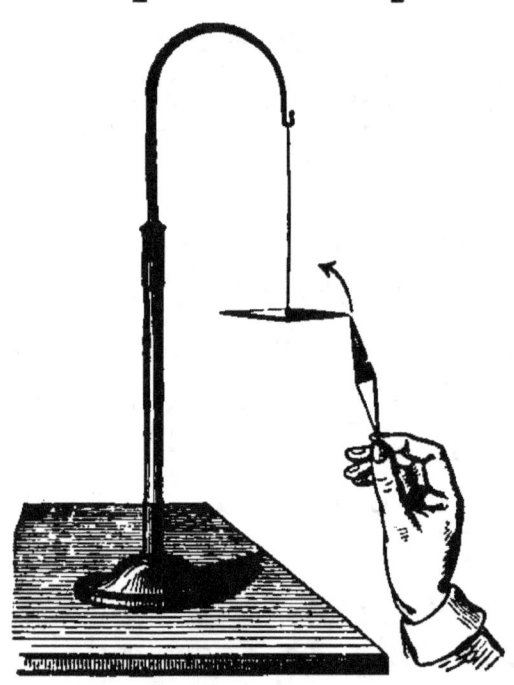

bringing the S.-seeking pole of the bar magnet near the N.-seeking pole of the needle: again there is strong *attraction.*   Reverse the bar magnet, and its N.-seeking pole as strongly *repels* the N.-seeking pole of the needle (see Fig. 15).   These experiments may be repeated with other bar magnets, with horse-shoe magnets, and with lode-stones, and in every case we find the same phenomena occur.

FIG. 15.—How a Magnet acts upon another Magnet: N. pole repels N. pole.

27. **Magnetic Repulsion.—**

North Pole repels North Pole.
Marked Pole repels Marked Pole.
Red Pole repels Red Pole.

So also

South Pole repels South Pole.
Unmarked Pole repels Unmarked Pole.
Blue Pole repels Blue Pole.

28. Magnetic Attraction.——

North Pole attracts South Pole.
Marked Pole attracts Unmarked Pole.
Red Pole attracts Blue Pole.

And also

South Pole attracts North Pole.
Unmarked Pole attracts Marked Pole.
Blue Pole attracts Red Pole.

29. Law of Magnetic Attraction and Repulsion.——
Putting together the results of all these experiments
on the action of the poles of magnets upon one
another, it is possible to establish a rule or law
which will always hold good, and which will express
the whole matter in a few words. The law is a
very short one, and is as follows :——

Like Poles repel,
Unlike Poles attract.

If this law is carefully borne in mind, we shall always be able to explain the action of one magnet
upon another magnet.

**30. Mutual Action of Poles as shown by Lines of
Force.——**We can employ Faraday's method of sprink-
ling iron filings upon a card, beneath which magnets
may be placed in various positions, to discover the
action of pole upon pole at a distance. First, let
two bar magnets be placed with their *unlike* poles
facing one another, say an inch apart : then the lines
of force run from pole to pole, showing that " unlike
poles attract " (Fig. 16, A). Now reverse one of the

magnets, so that *like* poles face each other: the curves from each pole refuse then to join with one another, but bend round and away, proving that "like poles repel" (Fig. 16, B).

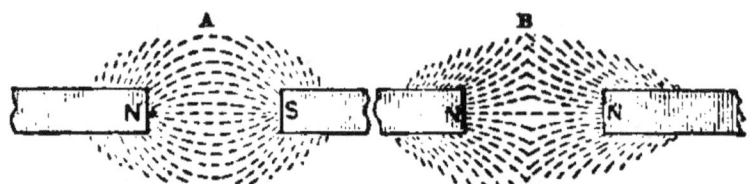

FIG. 16.—Curves of Force: like poles repel, unlike poles attract.

31. Diminution of Repulsion with Increase of Distance.—The force of magnetic repulsion is diminished as the distance is increased, according to the same law which governs the force of attraction. The force with which the pole of one magnet repels the like pole of another magnet diminishes *inversely as the square of the distance* between the two poles. Thus, at five inches apart the repulsive force will be only one-twenty-fifth as strong as at one inch; at a distance of ten inches the repulsion will be one-quarter of what it would be at a distance of five inches, and one-hundredth of that at one inch, and so on.

32. Unlike Poles neutralize one another.—When poles of equal strength but of opposite names are placed close together, their magnetic forces are spent upon each other, and they are unable to exert any influence upon magnetic substances. As an illustration, let a small iron key, or a nail, be suspended from one pole of a magnet; now slide the opposite pole of a second magnet of about the same strength along the upper surface of the first magnet (Fig. 17). When the unlike poles come in contact they neutralize one another, and the suspended piece of iron is then brought to the ground by the force of gravity.

33. Definition of Polarity.—We now see that a magnet is a *two-ended* substance—" two-ended " in the sense that the pro-

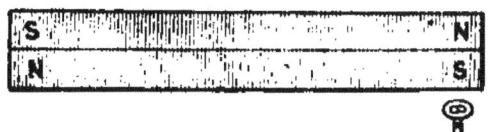

Fig. 17.—Slide one Magnet over another, and so cause key to drop.

perties of the one end are different from those of the other end. The N.-seeking pole of any magnet has properties distinct from and opposite to the properties possessed by the S.-seeking pole of the same magnet. We also find that it is the *opposite* properties which have an attraction for one another—north magnetism attracts south magnetism, and south magnetism attracts north magnetism also. But north magnetism repels—dislikes, as it were—north magnetism, and south magnetism repels south magnetism. Further, whatever amount of magnetic force there may be at one pole of a magnet, there will be a precisely equal amount at the other pole of the same magnet. If the N.-seeking pole of any magnet can just sustain a piece of soft iron weighing ten ounces, then we may be sure that that is also the greatest weight which can be sustained by the S.-seeking pole of the same magnet.

That the amount of north magnetism in any magnet is accurately equal to the amount of south magnetism in the same magnet may be beautifully shown by taking a magnet made of thin flexible steel and bending it until the two ends touch each other (Figs. 18 and 19). While the ends continue to touch, the magnet will have no power to attract the smallest piece of iron; the north magnetism at a and the south magnetism at b attract one another,

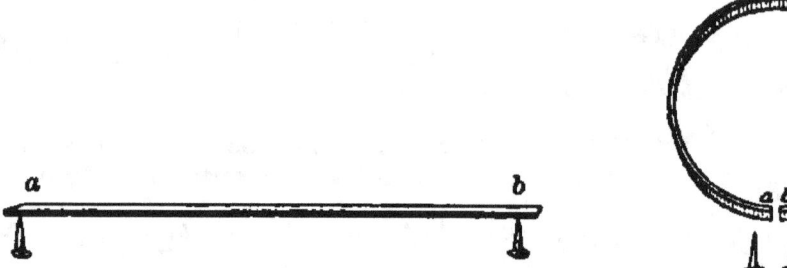

FIG. 18.—Thin straight steel Magnet supporting two nails.

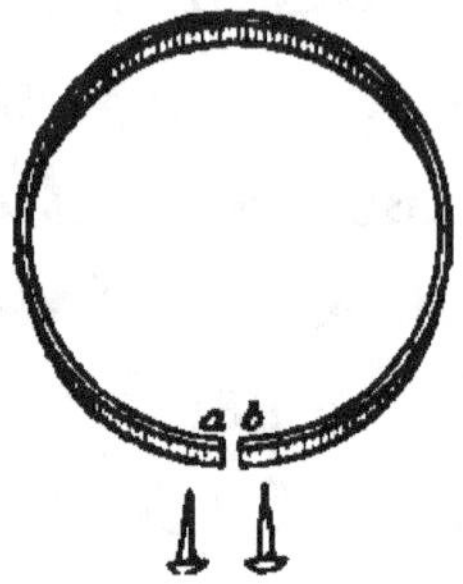

FIG. 19.—Magnet bent so that poles touch: nails then fall off.

and as they are exactly equal in quantity within the magnet, there is no magnetism of either kind left to exert any magnetic force on bodies outside the magnet. From the consideration of such experiments as these, the following definition of "polarity" has been drawn up :—" Opposite properties in opposite directions, so exactly equal as to accurately neutralize one another."

VI.—THEORIES OF MAGNETISM.

34. What is a "Theory"?—If we *see* a man knocked down by a runaway horse, we do not form any "theory" as to the cause of his fall; we see and know the *fact*. But if we find a man lying insensible on the road, we may *conceive* of several causes which may have produced this effect—as a blow, a fit, etc.—and each of these ideas would represent a distinct "theory" of the cause which led to the man's fall. Only one of these theories could be true, and a careful examination of the man's body by a doctor would probably show which was the *right* theory.

35. Magnetism is one of the Forces of Nature.—What can "magnetism" be? It is not any kind of *matter*, for it has no *weight*. A bar of steel weighs not the fraction of a grain heavier after it has been made an artificial magnet than it did before. Magnetism, then, must be a *force*. But how does this force act? What is the cause of the difference between the properties of the two poles of a magnet? Why is it that permanent magnets can only be made of hard steel ? Above all, how is it possible for a magnet to *act at a distance?* Several theories have been framed to account for all the facts which have been observed, but we can here only consider two—namely, (1) the two-fluid theory, and (2) the theory which assumes the polarity of the molecules.

36. Nature of Fluids.—The term " fluid " includes both liquids and gases. Air is a fluid which surrounds the Earth, reaching to a height of about two hundred miles. Beyond the air, it is believed that all the space, as far as the sun, moon, and stars, is occupied by a substance called the *ether*, which is so excessively fine, or "rare," that its weight cannot be detected by our most delicate balances. Not only does this ether fill all the space outside the Earth, but it also pervades all the matter of which the Earth is composed, mingling with every substance, and passing through the pores of solids " as easily as the wind blows through a grove of trees." It is impossible to obtain a space free from this ether. When we remove the air from a glass bottle by means of an air-pump, the interior of the bottle is only a *vacuum* in the sense of being nearly free from air ; it is still full of ether.

Those who believe in the two-fluid theory of magnetism consider that in every magnet, and in every magnetic substance, there are two fluids which resemble the ether in many of their properties, especially in being without any sensible weight.

37. Physical Composition of Matter.—In considering this two-fluid theory, we must understand the mean-

ings of certain words which are applied to portions
of matter. Any piece of matter of a tolerable size
—at least large enough to be handled—we term
a *body;* thus a lump of sugar, a ball, a chair, etc.,
are examples of *bodies.*

A body may be broken up, as when we pound
lump-sugar in a mortar into *particles;* these particles are too small to handle, but they are large
enough to be easily seen.

If, now, we take the particles of sugar and shake
them up in a glass of water they will disappear,
although the presence of the sugar in the water is
revealed by the sweet taste. The sugar disappears
because the water divides its particles into smaller
pieces still—into pieces called *molecules,* which are
so small as to be invisible even through the most
powerful microscopes.

38. Two-fluid Theory of Magnetism.—Now, a bar of
steel is a body. When we file the steel, we break it
up into particles; and each particle of steel is composed of a great number of molecules. When the
bar of steel is unmagnetized, the two magnetic
fluids are combined with one another all around
each molecule, just as the two gases, oxygen and
nitrogen, everywhere mingle together to form the
atmosphere which encircles the Earth. In their
combined state the two magnetic fluids neutralize
each other, and the body in which they are con-
tained exhibits *no magnetic properties.*

But when the steel is magnetized, the north magnetic fluid is drawn together, and collects around
the same half of each molecule, while the south

magnetic fluid is concentrated around the other half of each molecule (see Fig. 20).   The result of this will be that at one

FIG. 20.—Two-fluid Theory.
Magnetized steel bar.

end of the steel bar there will be left free north magnetic fluid, and at the other end free south magnetic fluid, ready to influence any other substance containing magnetic fluids which may be brought near.   It will be seen that this theory requires :—

(1.) The existence in every magnet of two fluids of distinct properties.

(2.) These fluids must be self-repulsive but mutually attractive.

(3.) The two fluids must exist around every molecule of the substance.

(4.) The fluids must be capable of *separation*.

This two-fluid theory is a convenient one, and if we assume the existence of the fluids, the theory is capable of explaining all the facts about magnetism. But we must be careful, if we adopt this theory, only to use it as a means of representing, grouping, and explaining our observations.   The fact is, that we do not yet know what magnetism is, but it is certain that it is *not* a fluid.   Yet it may be useful, in the meantime, to conceive of it as a fluid, and this may help us to discover what it really is.   If, however, we can discover any simpler way of accounting for the facts of magnetism, then it seems right that we should adopt it *instead* of the two-fluid theory.   Such a simpler theory we shall now proceed to describe.

39. The Molecular Polarity Theory.—By the two-fluid theory we proceed to *cause* every molecule of a magnet to be itself a perfect little magnet, by effecting a separation of the two magnetic fluids in some way or other. But since we have to assume the existence of the two fluids, is it not better to consider every molecule of a magnet, or of a magnetic substance, to be naturally a magnet, and for the present to remain contented with this, and not endeavour to further explain *what it is* which causes each molecule to be a magnet? We cannot even explain *why* a stone falls. We say that it is due to the force of gravity, and Newton discovered for us the laws according to which the force of gravity acts. But as to *what* that force is, or *why* it is able to make the stone fall, we are utterly ignorant. It is the same with the force of magnetism. We can *prove* that every molecule of a magnet is itself a perfect little magnet; but *why* it is a magnet, or *what* makes it a magnet, we cannot certainly tell.

In the present state of our knowledge, it is the simplest and easiest theory of magnetism to assume that each molecule has polarity—that is, is a magnet—and to proceed to explain the facts we observe by referring them to this explanation.

VII.—POLARITY OF THE MOLECULES.

40. Result of breaking a Magnet—41. All Magnets are composed of Smaller Magnets.

40. Result of breaking a Magnet.—Let us magnetize an ordinary sewing-needle, or any piece of thin steel. The magnet so produced will have two poles and a neutral line. Let us now break this magnet; it does not matter at what point we break it, but suppose we do so at the neutral line. What is the result? Have we succeeded in obtaining two pieces of steel, each containing *one* pole? Certainly not. Each half of the magnet is as perfect a magnet as the original whole magnet; each half possesses *two* poles and a magnetic equator (Fig. 21). Break each half again, and test the four pieces; each piece is a perfect magnet. If we continue breaking the pieces until they become as small as particles of dust, still each particle will be found to be a magnet. It is impossible to avoid the conclusion that if we could go on breaking up these particles until we had separated them into their molecules, that *every molecule would be a perfect magnet.* After this, some one may inquire, Suppose the molecules themselves were to be

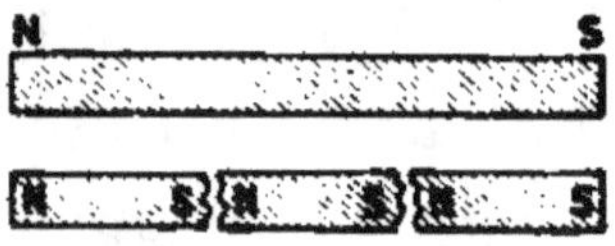

Fig. 21.—Result of breaking a Magnet.

broken, what would happen then ? To this we can only reply that it is impossible, by any physical means, to break a molecule.

41. All Magnets are composed of Smaller Magnets.— The experiment of breaking a magnet shows that a magnet does not consist of two halves, one containing nothing but north magnetism and the other nothing but south magnetism, but that there is both north and south magnetism in each half of any magnet. The experiment we are now about to describe may be considered as the exact opposite of the last one; it will show that by putting together a great number of small magnets it is possible to obtain a single large magnet. Take a glass tube and close one end with a cork; now fill the tube with *steel* filings, and cork up the other end. Proceed next to magnetize the tube full of filings, just as if it were an ordinary piece of steel, by rubbing it with a lodestone or with an artificial magnet, or, in fact, by any of the methods described in chapter viii. When this has been done, the whole tube will be found to behave exactly like a common bar magnet, and to have the usual two poles and one neutral line. Removing the cork from one end of the tube, pour out the filings and examine them: each tiny piece of steel will be found to be a magnet. Shake them up and replace them in the glass tube. Again test the tube, and it will be found to have lost its polarity, and therefore to be no longer a magnet. Again remove the steel filings from the tube, and carefully examine them; every particle of steel will be found, as before, to be a perfect little magnet. What, then, caused the loss of magnetic power which resulted from the shaking-up of the filings? When the mass of filings (each filing a tiny magnet) in the tube behaved like one large magnet, it was because all their N.-seeking poles pointed in one direction, and their S.-seeking poles

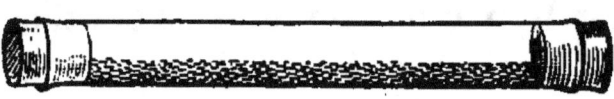

FIG. 22.—Glass tube with steel filings: magnetized.

in the other (Fig. 22). By emptying out and replacing the filings we destroyed this orderly arrangement of the poles, and the result was that, although each particle of steel

FIG. 23.—Glass tube with steel filings: after shaking up.

still remained a magnet, yet they did not act together; but, as the N.-seeking poles of, say, one-half of the filings pointed towards one end of the tube, while the N.-seeking poles of the other half pointed the opposite way, the little magnets neutralized each other, and the mass of filings, as a whole, showed no polarity (Fig. 23).

For exactly the same reason we find that a compound bar composed of four magnets of equal power, arranged so that two N.-seeking poles are at one end, while the other two N.-seeking poles are at the other end, has, as a whole, no magnetic force (Fig. 24).

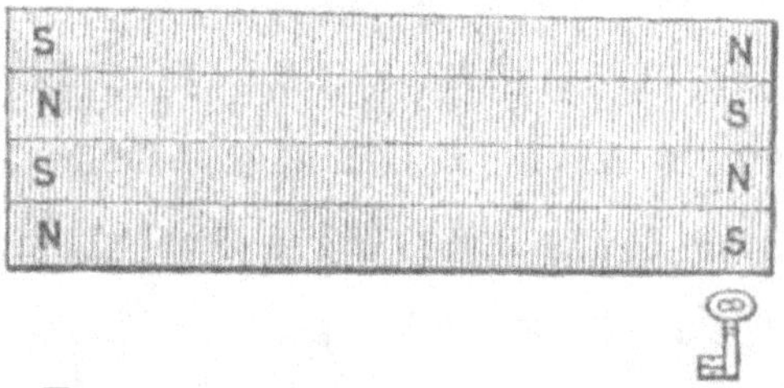

FIG. 24.—Compound bar; four equal magnets: no magnetic force.

This experiment strongly supports the theory that every molecule of a magnet is itself a perfect magnet.

VIII.—METHODS OF MAGNETIZATION.

42. How to make Permanent Magnets.—If we accept the theory of " polarity of the molecules," we can understand that in what we should call an " unmagnetized " bar of steel every molecule is really a magnet, but that, as there is no *arrangement* of the molecules—as their poles lie pointing in all directions—so the bar, as a whole, has none of the properties of a magnet. But if we can introduce an orderly arrangement among the molecules—if we can force them, or most of them, to lie so that all their N.-seeking poles point in one direction, and all their S.-seeking poles in the other, then the bar will exhibit magnetic properties.

According to the two-fluid theory, the act of magnetization will consist in the *separation* of the two magnetic fluids which exist in a state of combination around each molecule. When these fluids are separated, so that the north magnetic fluid is brought entirely to the same end of each molecule —leaving south magnetic fluid at the other end— then the whole bar becomes a magnet.

43. Method of Single Touch.—Take a piece of steel, a sewing-needle for instance, and stroke it several times, always in the *same direction* and with the *same pole* of a bar magnet (Fig. 25); stroke it also

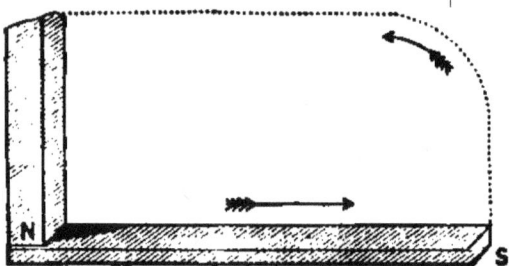

FIG. 25.—Method of Single Touch.

on each side. Of course, whether we stroke the needle with the magnet, or rub the magnet with the needle, will make no difference; the needle will quickly become a magnet. If we test the new magnet, we shall find that the end where the magnet was *taken away* from the steel is a pole of the *opposite name* to that with which it was rubbed. Thus, if we rub the needle from eye to point with a N.-seeking pole, then the point will become a S.-seeking pole; if we rub it the same way with a S.-seeking pole, then the point will be made a N.-seeking pole. We can imagine the north magnetism dragging the south magnetism after it, attracting the S.-seeking poles of all the molecules, and forcing them all to point in one direction. This method is usually employed only to magnetize *small* pieces of steel.

44. Method of Separate Touch.—Two magnets are required for this method, and their *opposite* poles must be employed. Commence in the centre of the bar to be magnetized, and *separate* the two magnets, moving one towards one end of the bar, and the other in the opposite direction. Lift up both the magnets when they reach the ends of the bar, and commence with them again in the middle (see Fig.

26). Repeat the process on each side of the bar, giving each side, say, half-a-dozen rubs. In this, as

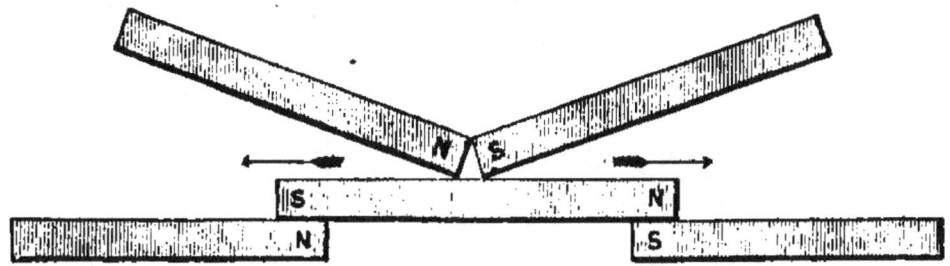

Fig. 26.—Separate Touch.

in the next method, it will be found advantageous to support the ends of the bar to be magnetized upon two magnets, arranged with their poles as in Fig. 26. The presence of these magnets enables the steel bar to be magnetized more rapidly and more effectually, for the attractive forces of their poles prevent the molecules of the bar from slipping round (on the molecular polarity theory), or the recombination of the two fluids (on the two-fluid theory), in the intervals during which the magnets are being raised and moved from the ends to the centre of the bar.

45. Method of Double Touch.—As in the last

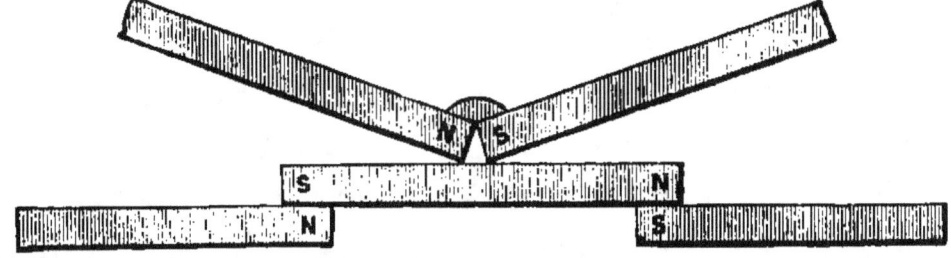

Fig. 27.—Double Touch.

method, use the opposite poles of two bar magnets, but place a piece of cork or of wood between them, to prevent the ends touching (Fig. 27). Commence in the *middle* of the steel bar to be magnetized, and move the two magnets *together*, first to one end of the bar, and then right back to the other end. Leave off in the middle, taking care to give each half of the bar the same number of rubs. Repeat this process on the opposite side of the steel bar, and, if the bar be of any thickness, on the other two sides also.

Permanent magnetization of steel by the electric current.—The most rapid and easy method of mag-

netizing ordinary steel bars is to pass them through a coil of silk-covered copper wire, through which a strong current of electricity is flowing (Fig. 28).

FIG. 28.—Steel bar passed through coil of wire carrying current.

Care must be taken always to pass the bar through the coil with *the same end first*. We must consider that the electric current causes the molecules of steel to place themselves so that their like poles all point in the same direction.

46. Consequent Points.—When the steel bar or rod is very long as compared with its thickness, as in the case of a knitting-needle, for example, or when it is incorrectly magnetized—either accidentally or on purpose—we may have four, six, eight, or more poles in the same bar of steel. In this case we may consider the bar as composed of two, three, or four magnets placed end to end, and *not* that we have one magnet with four or more poles. The additional poles situated at various parts of the bar are called

" consequent points," or " consequent poles " (Fig. 29). They may be readily discovered by the curves which will proceed from them if we lay a piece of

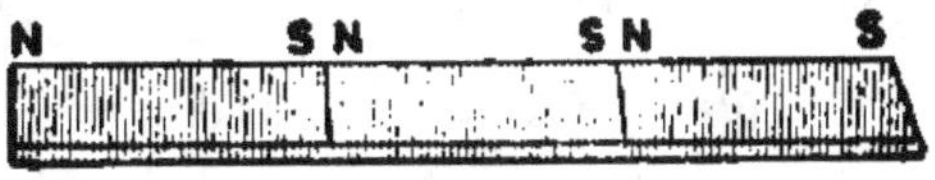

FIG. 29.—Consequent points or poles in steel magnet.

cardboard or glass upon the bar, and strew some iron filings upon it.

In this way it is possible to obtain a bar of steel which shall have a N.-seeking pole *at each end;* but we must not think that we have obtained a magnet with a single pole. Careful experiments will prove that in such a case there are two S.-seeking poles situated

FIG. 30.—Consequent points, with N pole at each end.

close together somewhere in the middle of the bar (Fig. 30).

47. Degree of Magnetization.—When a bar of steel has developed the full amount of magnetism which it is capable of *permanently retaining*, it is said to be *saturated.* The bar may be magnetized beyond this point, but the extra power so acquired is soon lost; in such a case we might say that the bar was *super-saturated.*

48. How to make an Electro-magnet.—Temporary, or electro-magnets, consist of pieces of soft iron, round which copper wire has been coiled (see Fig. 9). The wire must be covered with cotton, or, better, with silk, and it must be wound round the iron always in the same direction. The greater portion of the wire may be laid, coil upon coil, round the two *ends* of the soft iron bar, since no matter how strong a current of electricity we pass round the *middle* of the iron bar, we know that it can there produce no outward effect, for that is the position of the neutral line.   It has been found, by actual experiment, that when a very powerful electric current is passed round an iron bar (thus rendering it a magnet) the bar becomes a little *longer*. This is probably caused by the molecules turning round so as to set themselves parallel to the axis of the bar (see Fig. 32).

IX.—MAGNETIC INDUCTION.

49. Nature of Induction.—In the science of magnetism, and, indeed, in that of electricity also, we use the word "induction" to mean *action at a distance*.

How strange and mysterious is the manner in which a magnet can act across a vacant space upon another magnet, attracting or repelling it according as the poles nearest together are of unlike or of like kinds! If a piece of glass, or a sheet of paper, be placed between the two magnets, or between a magnet and a piece of soft iron, still the same "action at a distance" is perceived (Fig. 31). If, however, a sheet of *iron* be placed between the two magnets, then this substance acts as a screen or shield, and the

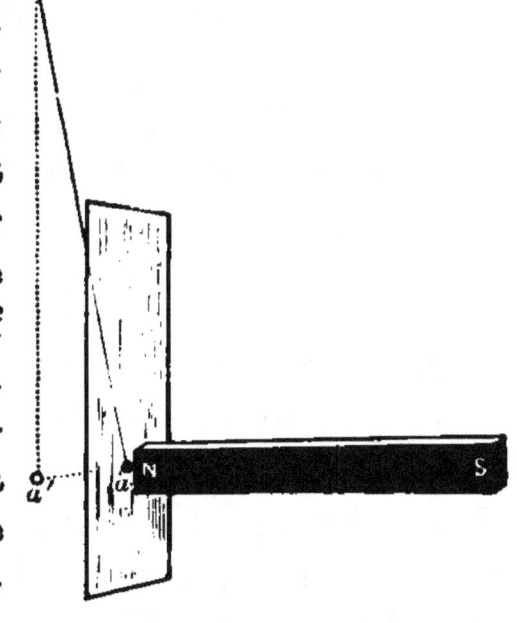

Fig. 31.—Magnet attracting iron bullet through sheet of glass.

one magnet is no longer able to influence the other, or to attract a piece of iron placed on the opposite side.   As induction takes place just as well through a vacuum as through the air, we are compelled to believe that the magnetic force is transmitted through space by the fluid called the *ether*, of which we have already spoken (see page 38).

50. Inductive Action of Magnets upon Soft Iron.— We can now, perhaps, conceive how it is that one magnet is able from a distance to affect another magnet: there is magnetic force in each, and this force is transmitted by the ether.   But what is it that enables a magnet to attract a piece of soft iron?   The answer to this is, that although we cannot make permanent magnets out of soft iron, yet a piece of soft iron can be made a *temporary* magnet with the greatest ease.   Let *a b* (Fig. 32)

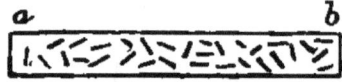

FIG. 32.—Bar of soft iron with molecules pointing in various directions.

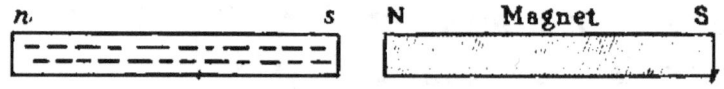

Same bar under Induction.

represent the state of the molecules of a bar of iron —a tenpenny nail, for instance—under ordinary conditions.   Each molecule of iron has polarity, but as they point in all directions, the bar, as a whole, has no magnetic power.   But let a magnet, N S, be brought within half-an-inch of one end of the iron bar; immediately this is done, it will be found that the iron has all the properties of a magnet, possessing two poles at *n* and *s*.

The pole of the magnet, N, has evidently been able to act inductively across the space, N s, and to so affect the molecules of the iron that they have rearranged themselves, with their S.-seeking poles all turned towards the N.-seeking pole of the magnet. Remove the magnet, and the molecules of the iron swing round again, attracting one another; the bar loses its polarity, and is no longer a magnet.

51. Attraction is always accompanied by Repulsion. —If a small iron bar be suspended by a thread, and one end of a bar magnet be brought near it, the iron will dart towards the magnet, and will adhere to it. In Fig. 33, let N s be again the magnet, and n s the soft iron magnetized by induction. Then the pole, N, will attract s, but will repel n. The reason why the iron moves towards the magnet is that the two attracting poles, N and s, are much *nearer* one to the other than the

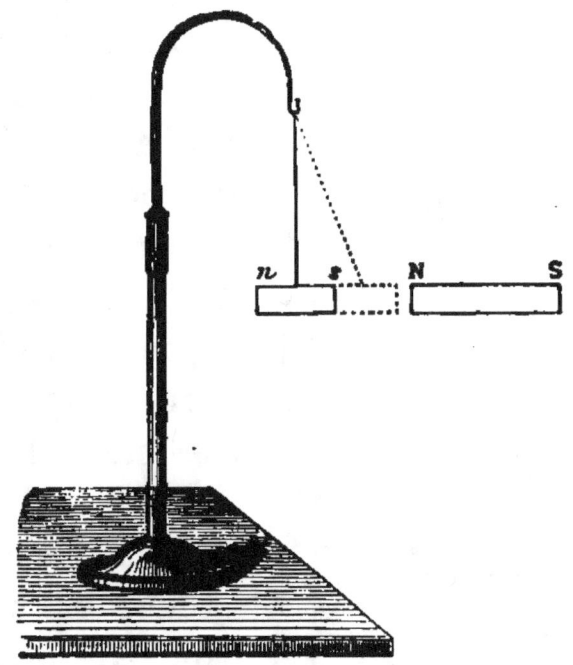

Fig. 33.—N s=1 inch. N n=2 inches. The soft iron, n s, moves towards magnet, for attraction equals four times repulsion.

two repelling poles, N and n. If N s be one inch, and N n be two inches, then the attraction will be four times greater than the repulsion (see page 29). Therefore any piece of soft iron moves towards a magnet; for although, while one of its ends is attracted the other is repelled, yet the attracting force is greater than the repelling force.

52. Magnets can only attract Magnets.—We are now able to understand the proverb, or saying, that "magnets can only attract magnets." The fact is, that before any piece of soft iron is attracted by a magnet, it is (temporarily) magnetized, and the magnet then acts upon it (while the iron reacts upon the magnet), just as one magnet acts upon another. To show that a piece of soft iron has all the properties of a magnet while it remains *near* a permanent magnet, we may suspend the small bar of soft iron, *n s*, above the powerful horse-shoe magnet, N s (Fig. 34). While the iron remains near the magnet it is itself a magnet, having poles at *n s*, the S.-seeking pole of the iron lying above the N.-seeking pole of the steel horse-shoe magnet, and the N.-seeking pole lying above the S.-seeking. To show the existence of these two poles in the iron bar, bring the N.-seeking pole of a bar magnet, N′ s′, near the end of the iron bar where the N.-seeking pole is supposed to lie; a repulsion of the end, *n*, will be obtained. In the same way the end, *s*, will be repelled by s′. Now remove the iron bar, *n s*, from its position near N s, and either end of it will now be attracted by either pole of the second magnet. The bar has thus lost its polarity, and is no longer a magnet.

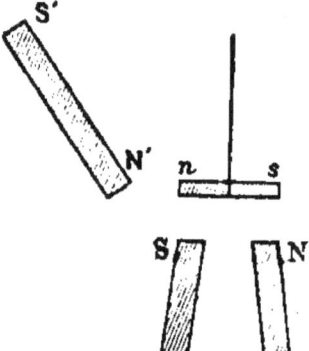

FIG. 34.—N s, horse-shoe magnet; *n s*, soft iron bar; N′ s′, bar magnet. *n s* being magnetized by induction, *n* is repelled by N′.

53. Chain of Magnets.—It is now plain that when a bit of soft iron (as $n\ s$ in Fig. 35) is allowed to touch one pole of a bar magnet it becomes magnetized. Its lower pole, s, is then able to attract a second piece of soft iron, $n'\ s'$, and

FIG. 35.—Chain of magnets.

this in turn a third piece, $n''\ s''$. In this way, with a powerful magnet, we may have a long chain of (temporary) magnets hanging from a pole. Now detach the uppermost bar, $n\ s$, from the magnet, and the whole chain falls to pieces. This must be one of the oldest experiments in magnetism, for it is described (see page 7) by the poet Lucretius, who lived two thousand years ago.

54. Repulsion as a consequence of Induction.—Suspend three or four pieces of iron wire from the same (say, the N.-seeking) pole of a magnet. The upper end of each piece of wire becomes, by induction, a S.-seeking pole. But the lower end of each wire is made a N.-seeking pole ; and since "like poles repel," these lower ends stand away one from the other, and the appearance is something like that of a fan or a brush (Fig. 36).

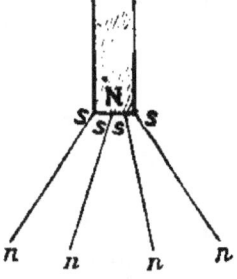

FIG. 36. — Iron wires hanging from one pole of bar magnet.

X.—MAGNETIC SUBSTANCES AND MAGNETS.

55. Magnetic Substances.—A distinction has been drawn between a *magnetic substance* and a *magnet.* The chief points in which they differ are shown in the following table :—

A MAGNET Has two poles.	A MAGNETIC SUBSTANCE Has no polarity.
Can be *repelled* as well as attracted by another magnet.	Is attracted by a magnet, but cannot be repelled.
Can either attract or repel another magnet.	Can attract a magnet, but cannot repel it.
Can attract all magnetic substances.	Cannot attract another magnetic substance.

56. How to distinguish Magnets.—The true test of a magnet is its power of *repulsion.* If we dip a steel bar into filings, and they adhere to it, that is not a certain proof that the bar is a magnet; for the filings themselves may be magnets, in which case they would adhere to a piece of soft iron if it were placed among them. And if, with the same steel bar, we hold one end of it near to the N. pole of a suspended magnet, and see the magnet move towards the bar, that, again, is no proof of the magnetism of the steel bar; for a piece of soft iron would do as much. But if we now present the same end of the steel bar to the S.-seeking pole of the magnet (Fig. 37), and see that pole *move away*

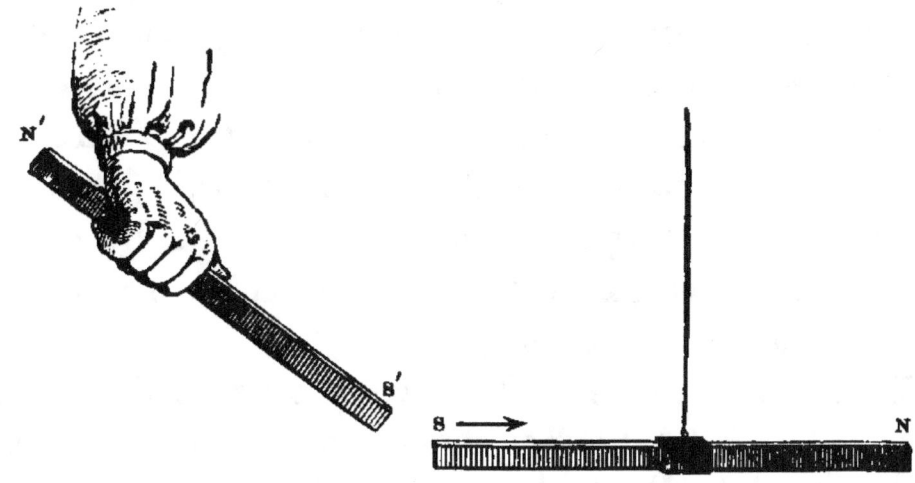

FIG. 37.—How to distinguish magnets.

from the bar, then we may be sure that our steel bar is a magnet, for it has *repelled* another magnet, and must therefore itself have polarity.

57. List of Magnetic Substances.—The most powerful magnets have no attractive power upon such substances as chalk, wood, glass, etc. Indeed, the number of magnetic substances, or those which are attracted by an ordinary magnet, is very small, as will be seen by the following list :—

Iron.	Cobalt.
Steel.	Manganese.
Nickel.	Chromium.

Of these, iron and steel stand far above the rest in magnetic power. The metals nickel and cobalt are also capable of a fair degree of magnetization. Two more metals, manganese and chromium, are feebly attracted; and there are a few other substances, including oxygen gas, which are very feebly attracted when tested with the most powerful electro-magnets. Bodies which are *attracted* by magnets are known as *paramagnetic*. It is probable that *all* substances are *influenced* in some way by the magnetic power; but we restrict the term " magnetic substances " to those which are *attracted* by a magnet. There are even some substances, such as the metals bismuth, antimony, and copper, which are *repelled* by either pole of a powerful magnet, and these are called *diamagnetic* substances.

58. Coercive Force.—How can we account for the fact that permanent magnets cannot be made of soft iron, but of hard steel only? It is soon found by experiment that soft iron is far easier to magnetize, temporarily, than steel. A suspended steel bullet, for example, is not attracted nearly so strongly as an iron bullet. The experiment shown in Fig. 35 will not succeed half so well if steel bars be substituted for the soft iron bars there employed to make the chain of magnets. There must then be some force in steel which *resists* magnetization, and to this the name of *coercive force* has been applied. The best hard steel has great coercive force; the best soft iron has little or no coercive force. We can conceive the coercive force to be that which *prevents* any movement of the molecules. In the soft iron the molecules swing round with great ease. The presence of a magnet near a piece of soft iron is sufficient to cause the molecules of the iron to arrange themselves so as to give polarity to the mass (see Fig. 32). In steel, the case is very different—the molecules are very hard to move. The pole of a magnet must be passed many times along each side of a steel bar before all the molecules can be made to place themselves endways, their like poles all pointing in the same direction. But when this is once done, the effect is permanent. It was difficult to turn the molecules ; but when the turning was effected, *they remained in their new position*, and the steel bar then possessed permanently the properties of a magnet. The greater the coercive force of any piece of steel, the more powerful will be the magnet into which it can be made.

59. Action of Keepers.—The hardest steel has not sufficient coercive force to keep its molecules always set in one direction. For this reason a single bar magnet will lose much of its magnetic power in a few weeks if it is left by itself. Bar magnets should be kept in pairs (see Fig. 3), the opposite poles being connected by pieces of soft iron called keepers. When a horse-shoe magnet is not in use, a similar keeper should be placed across its poles (Fig. 38). The soft iron becomes a magnet by induction, and then *reacts* upon the permanent magnet, preserving its magnetism, or even increasing it. *Three* magnets may be arranged in the form of a triangle, so that keepers may be dispensed with, care being taken to place *opposite*

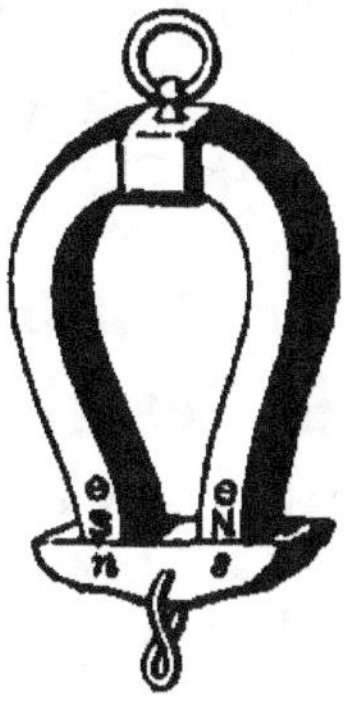

FIG. 38.
Horse-shoe magnet with keeper (latter = induced magnet).

poles one upon the other.   It will be better to place a piece of cardboard, or some other non-magnetic substance, between the poles.

60. Effect of Heat upon Magnetic Substances.—In Fig. 31, let a be a small iron ball suspended by a thin wire, and attracted towards the magnet, N.   Now heat the ball by means of a Bunsen burner, or in any other way.   As the ball becomes hot, it slowly moves away from the magnet, until, when it is red-hot, it hangs vertically as at a', the magnet having no power over it.   Remove the source of heat, and as the ball cools it will once more move towards the magnet.   From this we learn that magnetic substances lose their power of becoming temporarily magnetic while they are strongly heated.

61. Effect of Heat upon Magnets.—To test the effect of heat on magnets, we may magnetize a steel sewing-needle, and then place it in a clear fire and raise it to a bright red heat.   Allow the needle to cool, and test it, when it will be found to have lost its polarity and to be no longer a magnet.

The explanation of these effects of heat upon magnetic bodies and upon magnets is, that heat causes the molecules to vibrate rapidly, and therefore disturbs their arrangement as a whole.   To act as a magnet, either temporary or permanent, the molecules must rest with their like poles pointing steadily in one direction; but the force of heat interferes with this, for it sets the molecules in rapid motion, and causes them to point in various directions.

XI.—THE EARTH A GREAT MAGNET.

62. Why a Magnet points to the North.—In chapter I. we learned that although the first fact which attracted notice to magnets was the discovery of their power to attract iron, yet that it was afterwards found out that they also had the valuable property, when suspended or balanced so that they could freely move in any direction, of always pointing towards the Earth's North Pole. Ever since the time of this discovery, which in Europe was not made till about the year 1300, the magnetic needle, balanced on a pivot or point and protected by a case, has been of inestimable service to mankind under the name of the *compass*, guiding sailors across trackless seas, and directing travellers in unknown lands and during dark nights.

But *why* does a magnet point north and south ? The famous Dr. Gilbert, physician to Queen Elizabeth, and a most earnest experimenter in magnetism, made the great discovery (about the year 1600) that a magnet points north and south *because the Earth is a great magnet,* having its two magnetic poles situated not far from the geographical or true North and South Poles.

63. The Common Compass.—The ordinary compass sold by opticians consists of a small square or circular box, inside which (on the bottom of the box) is a card marked N., S., E., W., etc. The magnetic needle (Fig. 39) is balanced on an upright pivot in the centre of the card;

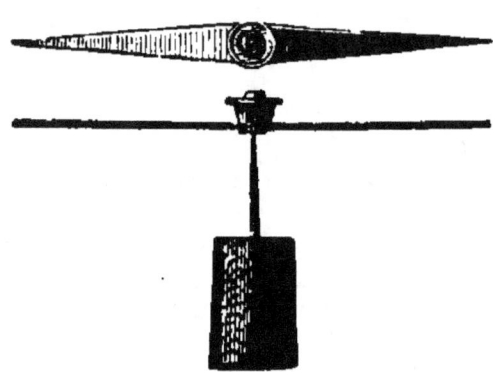

FIG. 39.—Needle of Compass.

thus the card is fixed, while the magnet is movable.

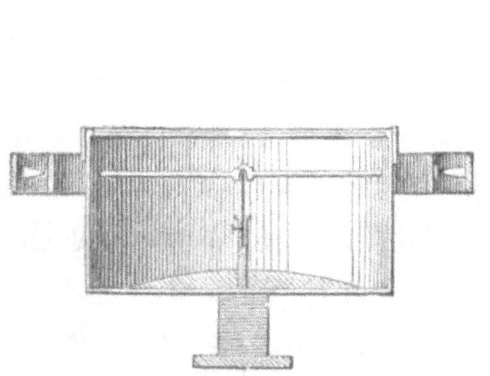

FIG. 40.—Section of Mariner's Compass.

FIG. 41.—Compass Card and Rings (Gimbals).

64. The Mariner's Compass.—The compass used by mariners, or seamen, has the magnetic needle *fixed* to the under side of a circular card; thus the card and the needle *turn together* on the pivot which rises from the bottom of the compass-box (Fig. 40), and the point marked N. on the card always points to the north.   Looking at the mariner's compass from the outside (Fig. 41), the magnetic needle is invisible, since it lies *underneath* the card. The card is divided into the "thirty-two points of the compass" (Fig. 42), of which the four principal (N., S., E., and W.) are called the cardinal points. Commencing at the north, and passing to the east or to the

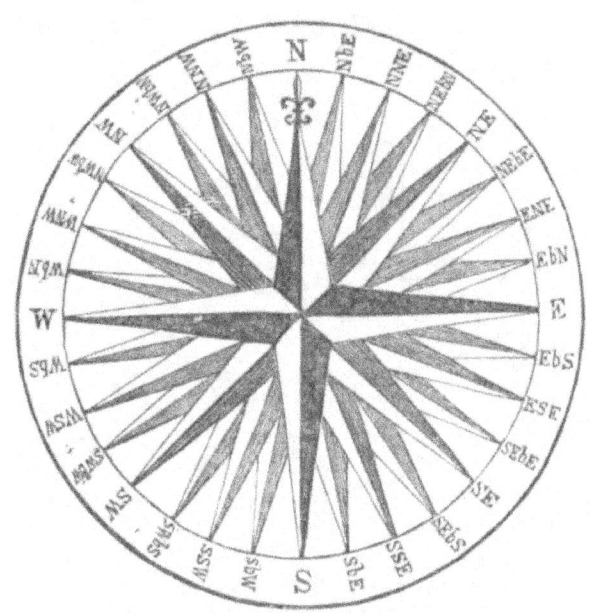

FIG. 42.—Points of the Compass.

west, the thirty-two points are named as follows:—

POINTS OF THE COMPASS.

North.
North by east.
North-north-east.
North-east by north.

North-east.
North-east by east.
East-north-east.
East by north.

East.
East by south.
East-south-east.
South-east by east.

South-east.
South-east by south.
South-south-east.
South by east.

South.

North.
North by west.
North-north-west.
North-west by north.

North-west.
North-west by west.
West-north-west.
West by north.

West.
West by south.
West-south-west.
South-west by west.

South-west.
South-west by south.
South-south-west.
South by west.

South.

65. Position of the Compass in a Ship.——In order to

keep the needle and card horizontal, notwithstanding the rolling or pitching of the ship, the compass-box is suspended by two copper rings, called gimbals (Figs. 41 and 43), and the whole is enclosed in a conical box named the binnacle, which is placed just in front of the helmsman, so that he can see exactly in what course or direction to steer the

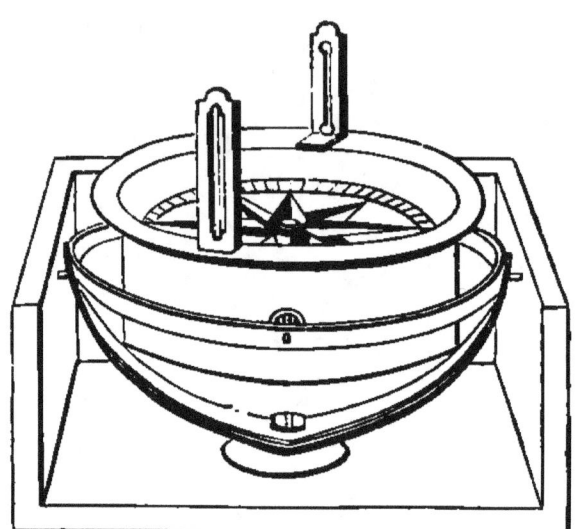

FIG. 43.—Mariner's Compass.

ship (Fig. 44). The mariner's compass is, of course, affected by the presence of masses of iron; and on board iron ships, or on wooden ships carrying cargoes of iron, the result has many times been to cause the ship to be steered on a wrong course, and so, perhaps, to be wrecked. To prevent this, it is now customary to carry a second compass, which is

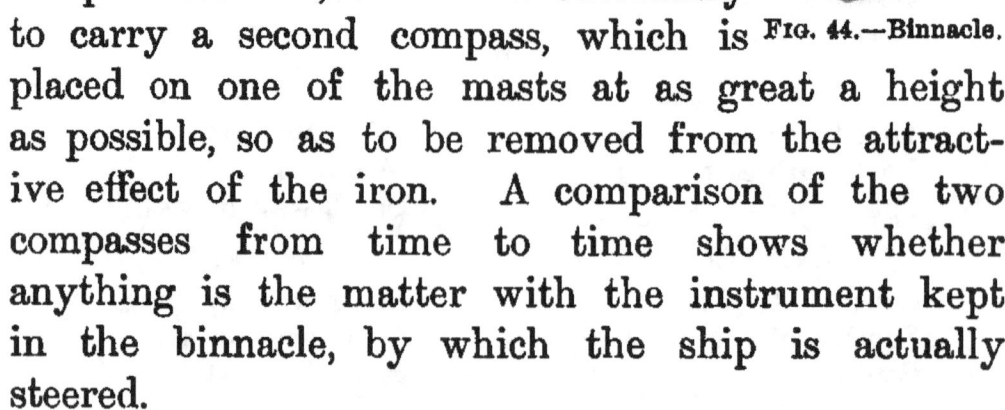

FIG. 44.—Binnacle.

placed on one of the masts at as great a height as possible, so as to be removed from the attractive effect of the iron. A comparison of the two compasses from time to time shows whether anything is the matter with the instrument kept in the binnacle, by which the ship is actually steered.

66. Magnetic Declination.—The geographical North and South Poles of the Earth are the ends of its axis of rotation: the Pole Star stands vertically over the Earth's true North Pole.

Now, in England, a magnet does not point to the geographical North Pole, but in a direction (in the centre of England) about eighteen degrees west of true north (Fig. 45). This *angle* between the magnetic meridian and the geographical meridian (or line of longitude) is called the *declination* of the magnet (Latin, *de*, from, and *clino*, I bend), to express the fact that

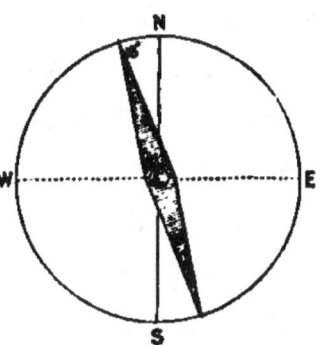

FIG. 45.—Declination.

a magnet declines from or deviates from a true northerly line. Sailors call it the *variation*, and they say that the compass-needle " varies " from true north.

67. The Declination is different in different Countries. —If we were to take a voyage to America, we

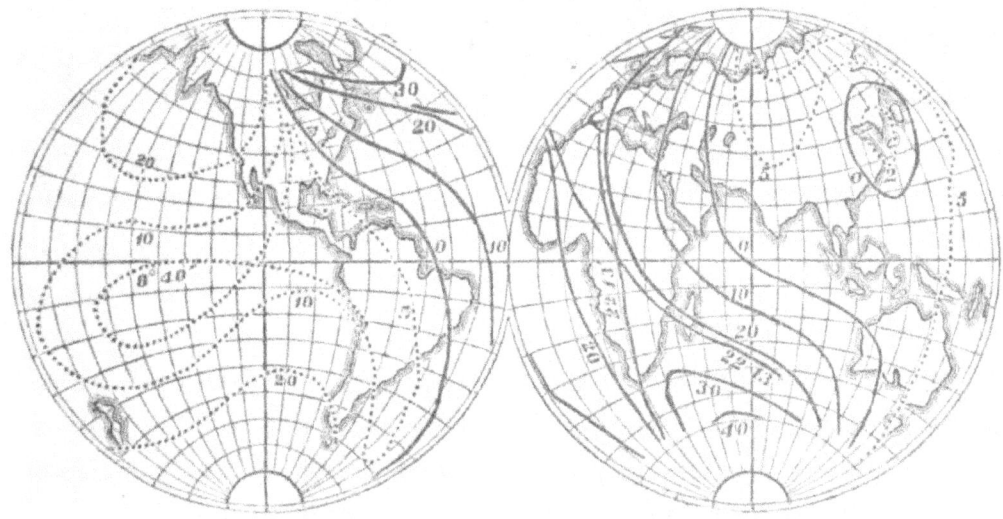

FIG. 46.—Lines of Equal Declination.

should find the declination to decrease as we advanced westward, until at last, near New York. we reached a line of no declination, where the compass-needle would point true north. Going west of this line, we should find the declination to be *easterly;* so that, at San Francisco, the magnet points at about the same angle to the *east* of true north as it does to the *west* in England.

68. The Declination varies, at the same Place, from year to year.—Although, in London, the compass-needle now points about 17° west of true north, yet in 1580, the first year in which accurate observations were made, it pointed (at the same place) $11\frac{1}{4}°$ *east* of true north. In 1657, the declination at London was 0°—that is, the needle there pointed due north. After this date the needle each year pointed more and more westward, until the extreme western declination ($24\frac{1}{2}°$) was reached in 1816. Since that date the declination has diminished yearly, until now it amounts to about 17° only. Similar changes have been observed at all localities where careful observations have been made.

XII.—TERRESTRIAL MAGNETISM.

69. Position of the Earth's Magnetic Poles.—We can easily understand the cause of magnetic declination. The Earth's magnetic poles are situated at some distance from its geographical poles, and therefore the magnetic meridians (see Fig. 46), which meet in the magnetic poles, must, as a rule, cut across the geographical meridians, or lines of longitude, which meet in the true North and South Poles.

Wherever we place a magnet upon the Earth, it will point to the magnetic poles, so that by marking upon a globe the lines along which the magnets point in different countries we shall be able to find, upon the globe, the spots where all these lines would meet, and at or near those spots will be the Earth's magnetic poles.

But the distribution of the Earth's magnetism is not so regular that we can be sure of the *exact* position of the Magnetic North and South Poles by this method. The North Magnetic Pole was, however, reached by the famous navigator, Sir James C. Ross, in 1831. He found it to be then situated in

Boothia, a peninsula in the extreme north of North America. Its latitude was 70° north, and its longitude 96¾° west. It was thus twenty degrees of latitude, or 1,400 miles, distant from the true North Pole.

The South Magnetic Pole has never been reached, for it lies within that great region of ice which encircles the geographical South Pole; but by *calculation* it is believed to be in south latitude 75°, and east longitude 154°, lying about 2,500 miles south of Australia, and about 1,000 miles from the true South Pole.

70. Magnetic Inclination, or Dip.—In 1576 a London instrument-maker, named Norman, discovered that if a steel needle (Fig. 47) is carefully balanced so that it lies horizontally, and is then magnetized and balanced from the same point as before, that its N.-seeking pole will slant or *dip* downwards. He constructed a piece of apparatus, termed a dipping-needle (Fig. 48), to measure the amount of the dip or inclination, and found that it then amounted, at London, to nearly 72° from the horizontal line. The dip continued to increase in amount until 1720, when it was 74¾° at London. It has since steadily diminished, and is now (in central England) 67½°. To show the dip, we may

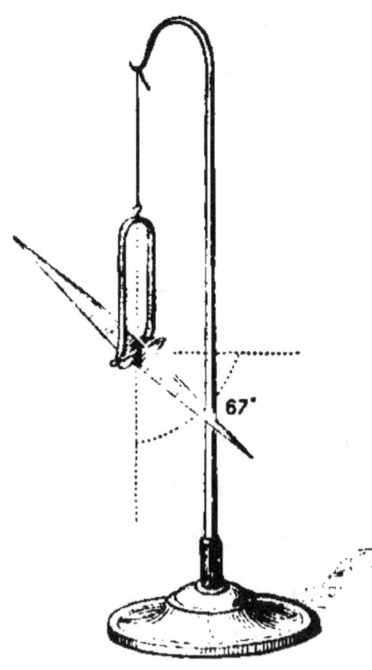

FIG. 47.—Simple Dipping-needle.

balance a steel knitting-needle upon the edge of a knife, and mark, with a spot of ink, the place where it must be supported to lie horizontally (Fig. 49). By means of a little hot shellac we fasten a fibre of silk to the marked spot, and again notice that the unmagnetized needle hangs horizontally when suspended. Now magnetize the needle, and suspend it as before. The end containing the N.-seeking pole will then be found to dip down at a considerable angle (Fig. 50).

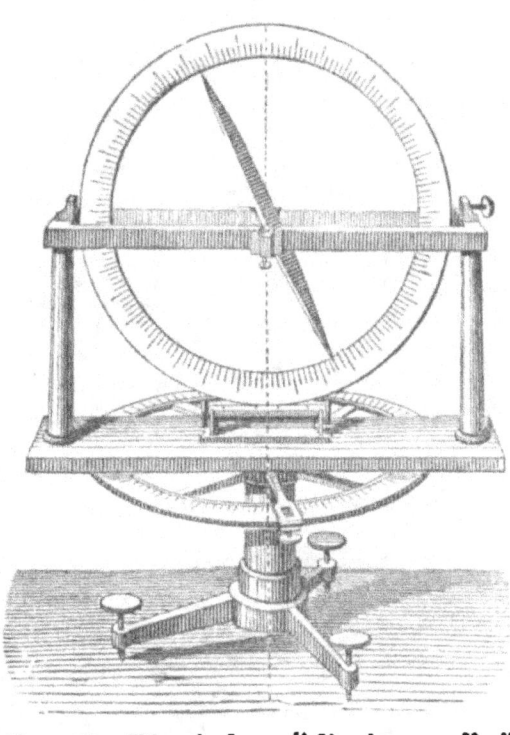

FIG. 48.—Dip-circle or "dipping-needle," for accurate observations of Inclination.

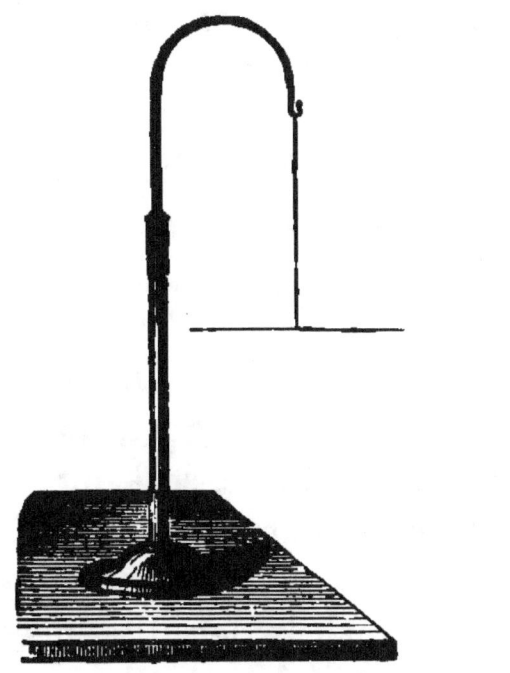
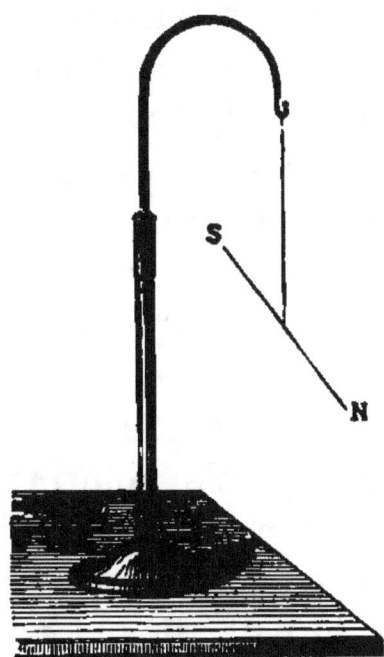

FIGS. 49, 50.—Knitting-needle before and after magnetization.

71. Amount of Dip at different Places on the Earth's Surface.—If we carry a dipping-needle from London towards the North Magnetic Pole, the dip will steadily *increase*, until, when Boothia is reached,

FIG. 51.—Needle carried from N. to S. Pole, to show Dip.

the needle will " stand on its head "—that is, will be vertical, dipping 90°. But if we travel southward the dip will gradually *decrease*, until we reach the Earth's *Magnetic Equator*, when the magnet will lie horizontally, having no dip. Continuing our southward journey, the S.-seeking pole of the needle will now be found to dip, until, could we reach the South Magnetic Pole, the dip would again amount to 90° (Fig. 51).

This behaviour of a dipping-needle on the Earth's surface may be imitated exactly by moving a small needle along a power-ful bar magnet, the needle being kept an inch or two above the magnet (Fig. 52).

FIG. 52.—Needle passed along above bar magnet, to show Dip.

When the needle lies over the neutral line, it sets itself horizontally, parallel to the magnet, its N.-seeking pole pointing towards the S.-seeking pole of the magnet. As the needle is steadily moved in the direction of the bar magnet's north pole, its own south pole dips downward, until, when the needle is directly over that north pole, the needle stands upright. Returning toward the neutral line, and passing the needle in the opposite direction, its north pole inclines, until, in turn, it dips 90° when the needle is directly over the S.-seeking pole of the bar magnet.

72. The Terrestrial Magnetic Equator.—Since the Earth has two magnetic poles, it must also have a neutral line, or Magnetic Equator, which can be dis-

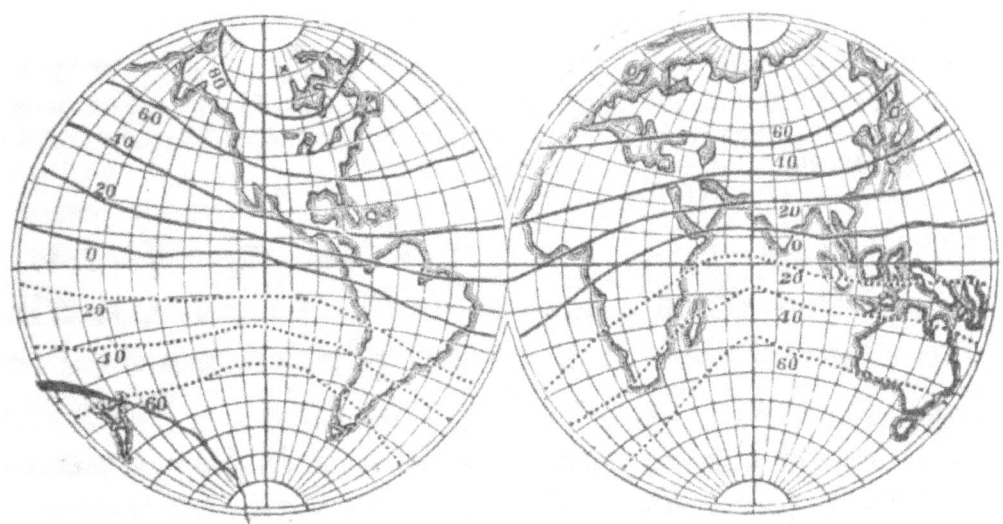

Fig. 53.—Lines of Equal Inclination or Dip, showing Magnetic Equator.

covered by the dip there being 0°, the needle lying horizontally. The Magnetic Equator is an irregular curved line surrounding the Earth in the same manner as the geographical Equator does, and coinciding, roughly, with the latter, which it crosses in at least four places (see Fig. 53). Lima and Bahia (in South America), the mouth of the river Niger (in Africa), Aden (in Arabia), and Madras, may be named as places in the vicinity of the Earth's Magnetic Equator.

XIII.—ACTION OF THE EARTH UPON MAGNETS.

73. Names of the Poles of a Magnet.—We have learned that when a magnet is free to move, it will place itself always in a certain position, pointing to magnetic north and to magnetic south. We have also learned that a magnet acts in this way because the Earth is a great magnet.

In ordinary language, people in the British Isles speak of that end of a magnet which points towards the Earth's North Magnetic Pole as the *north pole of the magnet.*

Students of magnetism will perceive that this is incorrect, since "*like* poles repel, while it is *unlike* poles which attract one another." In France, the end of the compass-needle which points northward is called the "*pole austral*"—that is, the south pole; while the end which points southwards is named the "*pole boreal*"—that is, the north pole. In this book we have called the ends of the magnet the "*north-seeking*" and "*south-seeking*" poles respectively, and to this there can be no objection; for, although that pole of a magnet which points northward is really the *south* pole of the magnet (taking the Earth as the standard magnet), yet it does *seek* to point to the north; so, also, it is true that the real north pole of a compass-needle is a *south-seeking pole*, for it always points to the Earth's south magnetic pole.

74. Magnetic Axis of a Magnet.—The axis of a magnet is the straight line drawn from one of its poles to the other. In a bar magnet the magnetic axis runs from end to end of the bar, through the centre; in a horse-shoe magnet, the axis runs straight across from pole to pole, in the direction of the keeper.

75. Position of Equilibrium of a Suspended Magnet. —When a body is at rest, it is said to be in equilibrium. Now, when a magnet is placed so that it is perfectly free to move in any direction, it will only remain at rest when it is lying in the *magnetic meridian*. The magnetic meridian may thus be considered as an extension of the magnetic axis of the magnet. If the suspended magnet be made to point east and west, it will, when released, instantly swing round to its old position, pointing to magnetic north and south.

The manner in which the Earth's magnetic poles act upon a magnet will be understood from an examination of Fig. 54. Here N and S are the terrestrial magnetic poles, while *s* and *n* are the poles of a magnetic needle. Two forces are at work upon the needle, and the directions in which these forces act are indicated by lines.   The pole N attracts *s*, but repels *n*; the pole S attracts *n*, but repels *s*.

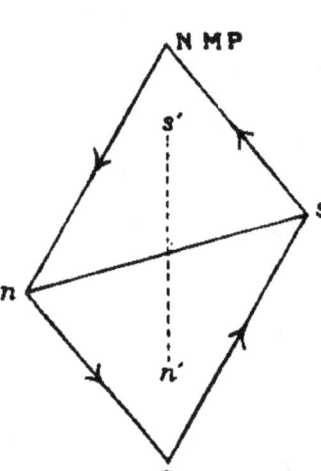

Thus there are two lines, N *s* and S *n*, representing an attracting force; while the lines N *n* and S *s* represent a repelling force. It will be seen that the two forces *act together*, and that the result of their joint action will be to cause the needle to swing round into the position *s' n'*.   So long as it remains in this position (shown by dotted lines) the magnet will be in *equilibrium*; but when the needle is forced aside, the magnetic forces residing in the Earth will endeavour to bring its axis back to the magnetic meridian.   When a system of forces acts in this way upon a body, it is called a *couple*; and the forces here represented constitute the terrestrial magnetic couple.

FIG. 54.—The Magnetic Couple.

76. The Earth's Influence on a Magnet is Directive only.—Let a thin, light magnetic needle be carefully balanced on a piece of cork, and placed in the middle of a large basin full of water.   If the magnet is placed so as to point east and west, it will, when released, swing round into the magnetic meridian, but it *will not move towards the Magnetic North Pole*; in fact, the Earth's influence is *directive* only. But if a magnet be brought to the side of the basin, the magnetic needle balanced upon the cork will move towards, or away from, the neighbouring magnet, according to the name of the pole which is presented to its nearest pole. Why then does not the magnet, when supported upon the surface of the water, move towards the Earth's North Magnetic Pole ?

77. Action of the Terrestrial Magnetic Couple.—The reason why the Earth's magnetism is unable to produce any motion of translation is to be found (1) in the *great distance* of the North Magnetic Pole from the balanced needle, and (2) in the fact that there is a force of repulsion at work as well as a force of attraction. The North Magnetic Pole attracts the N.-seeking pole of the needle, but it *repels* its S.-seeking pole. Now, supposing the experiment to be tried in England, and the needle to be three inches long, the distance from the North Magnetic Pole to the N.-seeking pole of the needle is about five thousand miles, and the distance of the S.-seeking pole of the same needle from the North Magnetic Pole will be five thousand miles and three inches. In so great a distance the difference—only three inches—between these two lengths is practically nothing, so that the S.-seeking pole is repelled *as strongly* as the N.-seeking pole is attracted. Therefore the needle, as a whole, does *not* move towards the North Magnetic Pole, but is simply swung round by the "terrestrial magnetic couple," until it is directed into the magnetic meridian.

78. Astatic Needles.—It is possible to so combine two magnetic needles that, while they each retain their magnetic properties, the combination will point indifferently in any direction. To such a couple of needles we give the name of an *astatic* pair, the word " astatic " expressing the fact that such a combination of magnets will not stand, or remain at rest, in any one position rather than another. An ordinary magnet, when free to move, will stand in a north and south direction only; an astatic pair will remain at rest if it is caused to point east and west, or in any other direction in which we may choose to place it.

To construct an astatic combination of magnets, we require two magnets of exactly equal power; and this equality it is very difficult to secure. Equality of strength may sometimes be obtained by carefully magnetizing a long steel needle, and then breaking it exactly in the centre. The two magnetic needles must now be fastened together, by a piece of copper wire for example, so that they lie with opposite poles one above the other, about an inch apart (Fig. 55), and the whole must be suspended by a fine thread, or balanced on a pivot. It is plain that if the two magnets are equal in strength, the magnetic

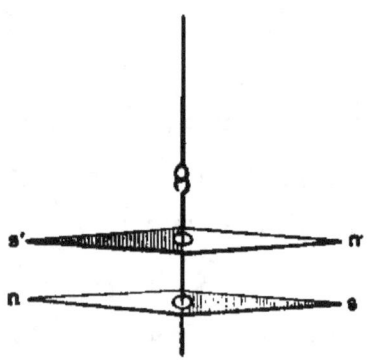

Fig. 55.—Astatic Combination of two magnetic needles.

poles of the Earth can have no power over such a pair of needles; for one needle cannot turn without the other, and while the North Magnetic Pole attracts *s*, it also repels *n'*, and while the South Magnetic Pole attracts *n*, it equally repels *s'*. Thus the pair of magnets, *n s* and *n' s'*, will remain pointing in any direction in which they may be left. The *use* of such a combination in delicate experiments will be plain when we consider that before any magnetic influence can produce motion in a *single* magnet, it has first to overcome the directive force of the Earth's magnetism. In an astatic combination the effect of the Earth's magnetism is neutralized, and the needles are left free to obey any other magnetic force, which may therefore be extremely small and yet produce a visible effect.

XIV.—INDUCTIVE ACTION OF THE EARTH.

79. Magnetization by Induction.—A very powerful magnet can magnetize a piece of steel without actual contact, although of course the magnets so produced are not so strong as they would have been if they had actually been rubbed by the magnet. For this reason it is wise to lay one's watch aside before experimenting with a large electro-magnet. All ordinary watches contain steel axles which can be magnetized by induction, and which then, acting magnetically upon the other iron and steel parts of the watch, hinder its action and prevent it from keeping good time. One way to prevent this would be to enclose the watch in an iron case, which would act as a screen. But this would be rather clumsy, so that watches are now made without any steel parts in their interior, for the use of electrical engineers and others who have much to do with powerful magnets; these are sold as " unmagnetizable " watches.

80. Magnetization by the Action of the Earth.—An examination of steel fire-irons, of iron railings, of lamp-posts, or of almost any object made of iron which has long remained in a vertical position, will show that such objects possess a feeble permanent magnetism. This magnetism has been produced by the inductive action of the Earth. The most favourable position in which to place a magnetic substance, in order that the Earth's magnetism may act upon it, will clearly be in the magnetic meridian, and slanting at the proper angle of dip. Accordingly, if in England we place a bar of soft iron so that it points a little (18°) west of true north, and has its north-pointing end inclined at an angle of 67½° from the horizontal line, we shall find that so long as the soft iron bar remains in this position *it is a magnet.* Even if the bar be merely placed upright, it will be found to possess polarity. This may be shown by passing a pivoted magnetic needle up and down the bar. The lower end repels the N.-seeking pole of the needle (therefore the lower end of the bar is a N.-seeking pole); as we pass the middle of the bar the needle swings round, and the upper end of the bar repels the S.-seeking pole. If the bar be now removed, and placed horizontally, it will be found to have lost all its (temporary) magnetism. A poker will answer for this experiment if it be made of fairly good soft iron; but with a *steel* poker no temporary magnetism can be thus obtained.

81. How to change the Temporary into Permanent Magnetism.——But if while the iron bar is vertical, or, better, while it is pointing (at the proper declination and dip) to the North Magnetic Pole, it be smartly and repeatedly struck with a hammer, then it will be found to *retain* its feeble magnetism, no matter in what position it may be placed.

Performing the experiment with a piece of iron wire instead of a bar, it will be found that forcibly *twisting* the wire by means of a pair of pincers will produce a similar effect. The hammering, or the twisting, serves to fix the molecules in the direction into which they have been brought by the inductive action of the Earth's magnetism. When the iron bar is *not* hammered, or the wire *not* twisted, then the molecules of iron swing back into their old diverse positions as soon as the bar, or the wire, is laid down.

82. Production of Natural Magnets.—We can now explain the origin of the magnetism of the lodestone, and why some beds of " magnetic iron ore " (Fe_3O_4) possess magnetic polarity, while others do not. When a layer of the ore lies vertically in the rocks, or, better still, if it happens to run in the magnetic meridian and at the angle of dip, then in the course of time it becomes magnetized by the inductive action of the Earth. But this will seldom be the case, and for this reason natural magnets are rare.

83. Motion of the Magnetic Poles.—The changes in the declination and in the inclination of the magnet observed at any given place can be explained by supposing a slow motion of the magnetic poles of the Earth. It would seem that the North Magnetic Pole revolves round the true North Pole in a period of about eight hundred years, and at a distance from it of about twenty degrees ; in a similar way the South Magnetic Pole slowly moves round the true South Pole. What the *cause* of these motions may be we do not yet know.

84. Origin of the Earth's Magnetism.—*Why* is the Earth a magnet? We must remember that there is a great abundance of magnetic substances in the crust of the Earth—a great quantity of iron ore, for example—and probably, at greater depths than our mines can attain, there is a great deal of metallic iron. We have also seen that it is possible, by means of a current of electricity, to produce the most powerful magnets. At present, the most probable theory as to the cause of the Earth's magnetic condition is that it is due to currents of electricity circulating round the Earth from east to west, principally in tropical regions and roughly parallel to the Equator. If we adopt this theory, we must find out some cause able to produce such electric currents, and this is to be found in the *heat* of the sun. It is a common experiment in the science of electricity to produce an electric current by the aid of heat. Now, as the Earth rotates on its axis from west to east, a flood of heat travels round it in the contrary direction, from east to west. This flow of heat produces currents of electricity, by which the Earth is magnetized. On this theory the Earth is a great electro-magnet. Many facts help to confirm this idea, for it is found that great disturbances of the sun's surface are usually followed by irregular motions of the magnetic needle. The light of the aurora, which circles in the air round the North and South Magnetic Poles, appears most frequently and is most brilliant when sun-spots are most frequent.

APPENDIX.

EDUCATION DEPARTMENT.

MAGNETISM.

New Code; Schedule IV.; Specific Subject, No. xii.—First Stage.

Syllabus for First Stage:—Attraction, Repulsion, and Polarity, as illustrated by the Magnet. Terrestrial Magnetism, and the Mariner's Compass.

QUESTIONS SET BY H.M. INSPECTORS OF SCHOOLS.

SET I.

1. What happens—
    (a) When a magnet is slightly heated?
    (b) When it is cooled again?
    (c) When a magnet is heated to redness?
    (d) When soft iron is heated to redness?

2. Give brief descriptions of the following :—
    An astatic needle.
    An electro-magnet.
    A compound magnet.

3. If a magnetized needle be suspended horizontally, how will it place itself in this country with respect to true north? Again, suppose a *truly* balanced magnetic needle be suspended so as to be free to move in any direction, how will it direct itself (a) at the magnetic equator, (b) at the magnetic poles? Give sketches.

SET II.

1. What is a magnet? How could you distinguish between a magnet and a piece of ordinary iron?

2. How could you show by experiment that not only does a magnet attract a piece of iron or steel, but also that the piece of iron or steel attracts the magnet?
3. How could iron filings be separated from sawdust by means of a magnet?
4. Describe the mariner's compass.

SET III.

1. How could you tell by means of a compass-needle whether a rod of steel is magnetized or not?
2. What are the *poles* of a magnet?
3. What is meant by saying that "the Earth is a magnet"?
4. I pick up a key by means of the north end of a magnet. What would be the effect if I slide gradually over this magnet the north end of a similar magnet.

SET IV.

1. Where are the magnetic poles of the Earth situated? Which of them has been visited, and by whom?
2. Explain the words *induction, repulsion, neutral,* and *molecule.*
3. Describe an experiment to show "Faraday's curves." How would you account for them?

SET V.

1. Name the cardinal points of the compass. How is it that the compass-card remains nearly horizontal, no matter how much a ship may roll or pitch?
2. Name all the magnetic substances you know. How could you distinguish a magnetic substance from a magnet?
3. How are magnets made by the method of "double touch"? Give a sketch.